U•X•L Encyclopedia of Biomes

Second Edition

U·X·L Encyclopedia of Biomes

Second Edition VOLUME 3

Marlene Weigel

U·X·L
A part of Gale, Cengage Learning

GALE
CENGAGE Learning

Detroit • New York • San Francisco • New Haven, Conn • Waterville, Maine • London

U·X·L Encyclopedia of Biomes
Marlene Weigel

Project Editor: Madeline Harris

Editorial: Kathleen Edgar, Debra Kirby, Kristine Krapp, Kimberley McGrath, and Lemma Shomali

Composition: Evi Abou-El-Seoud

Imaging: Lezlie Light

Manufacturing: Wendy Blurton

Product Design: Jennifer Wahi

Product Management: Julia Furtaw

Rights Acquisition and Management: Dean Dauphinais and Robyn Young

For product information and technology assistance, contact us at **Gale Customer Support, 1-800-877-4253.**
For permission to use material from this text or product, submit all requests online at **www.cengage.com/permissions.**
Further permissions questions can be emailed to **permissionrequest@cengage.com**

Cover photographs: Image copyright 2009: Debra Hughes (tree icon), Alexander Kolomietz (beach scene), Sergey Popov (baracudas), Snowleopard1 (frog), Chee-Onn Leong (tall pines), Sigen Photography (desert), and Oksana Perkins (pond and mountains), all used under licence from Shutterstock.com. Polar bear and fall leaves images © 2009/Getty Images.

Library of Congress Cataloging-in-Publication Data

U-X-L encyclopedia of biomes / [editor] Marlene Weigel. -- 2nd ed.
   p. cm. --
   Includes bibliographical references and index.
   ISBN 978-1-4144-5516-7 (set) -- ISBN 978-1-4144-5517-4 (vol. 1) -- ISBN978-1-4144-5518-1 (vol. 2) -- ISBN 978-1-4144-5519-8 (vol. 3) -- ISBN 978-1-4144-4426-0 (e-book)
   1. Biotic communities--Juvenile literature. I. Weigel, Marlene. II. Title: Encyclopedia of biomes.
QH541.14.U18 2009
577.8'2--dc22   2008014502

Gale
27500 Drake Rd.
Farmington Hills, MI, 48331-3535

978-1-4144-5516-7 (set)	1-4144-5516-X (set)
978-1-4144-5517-4 (vol. 1)	1-4144-5517-8 (vol. 1)
978-1-4144-5518-1 (vol. 2)	1-4144-5518-6 (vol. 2)
978-1-4144-5519-8 (vol. 3)	1-4144-5519-4 (vol. 3)

This title is also available as an e-book.
ISBN-13: 978-1-4144-4426-0 ISBN-10: 1-4144-4426-5
Contact your Gale, a part of Cengage Learning, sales representative for ordering information.

Printed in China by China Translation & Printing Services Limited
1 2 3 4 5 6 7 13 12 11 10 09

Table of Contents

Reader's Guide

This second edition of U•X•L Encyclopedia of Biomes offers readers comprehensive, easy-to-use, and current information on twelve of Earth's major biomes and their many component ecosystems. Arranged alphabetically across three volumes, each biome chapter includes: an overview; a description of how the biomes are formed; their climate; elevation; growing season; plants, animals, and endangered species; food webs; human culture; and economy. The information presented may be used in a variety of subject areas, such as biology, geography, anthropology, and current events. Each chapter includes a "spotlight" feature focusing on specific geographical areas related to the biome being discussed and concludes with a section composed of books, periodicals, Internet addresses, and environmental organizations for readers to conduct more extensive research.

Additional Features

Each volume of *U•X•L Encyclopedia of Biomes* includes color maps and at least 60 photos and illustrations pertaining to each biome, while sidebar boxes highlight fascinating facts and related information. All three volumes include a glossary, a bibliography, and a subject index covering all the subjects discussed in *U•X•L Encyclopedia of Biomes.*

Note

There are many different ways to describe certain aspects of a particular biome, and it would be impossible to include all of the classifications in *U•X•L Encyclopedia of Biomes.* However, in cases where more than one classification seemed useful, more than one was given. Please note that the classifications represented here may not be those preferred by all specialists in a particular field.

Every effort was made in this set to include the most accurate information pertaining to areas and other measurements. Great variations exist in the available data, however. Sometimes differences can be accounted for in terms of what was measured: the reported area of a lake, for example, may vary depending upon the point at which measuring began. Other differences may result from natural changes that took place between the time of one measurement and another. Further, other data may be questionable because reliable information has been difficult to obtain. This is particularly true for remote areas in developing countries, where funds for scientific research are lacking and non-native scientists may not be welcomed.

The U•X•L editors would like to thank author Marlene Weigel for her work on the original edition of this title. We also thank contributing writer Rita Travis for her work on the Coniferous Forest, Grassland, Tundra, and Wetland chapters. All the entries were updated by Lubnah Shomali for this new edition, and we relied heavily on our expert, Dr. Dan Skean, of Albion College to provide academic insight and additional feedback.

Comments and Suggestions

We welcome your comments on this work as well as your suggestions for topics to be featured in future editions of *U•X•L Encyclopedia of Biomes* Please write: Editors, *U•X•L Encyclopedia of Biomes,* U•X•L, 27500 Drake Rd., Farmington Hills, MI 48331-3535; call toll-free: 1-800-877-4253; fax: 248-699-8097; or send e-mail via www.gale.cengage.com.

Biomes of the World Map

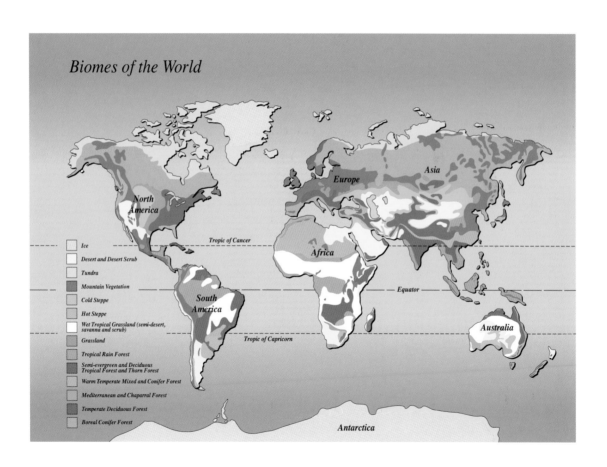

Words to Know

Abyssal plain: The flat midportion of the ocean floor that begins beyond the continental rise.

Acid rain: A mixture of water vapor and polluting compounds in the atmosphere that falls to Earth as rain or snow.

Active margin: A continental margin constantly being changed by earthquake and volcanic action.

Aerial roots: Plant roots that dangle in midair and absorb nutrients from their surroundings rather than from the soil.

Algae: Plantlike organisms that usually live in watery environments and depend upon photosynthesis for food.

Algal blooms: Sudden increases in the growth of algae on the ocean's surface.

Alluvial fan: A fan-shaped area created when a river or stream flows downhill, depositing sediment into a broader base that spreads outward.

Amphibians: Animals that spend part, if not most, of their lives in water.

Amphibious: Able to live on land or in water.

Angiosperms: Trees that bear flowers and produce their seeds inside a fruit; deciduous and rain forest trees are usually angiosperms.

Annuals: Plants that live for only one year or one growing season.

Aquatic: Having to do with water.

Aquifer: Rock beneath Earth's surface in which groundwater is stored.

Arachnids: Class of animals that includes spiders and scorpions.

Arctic tundra: Tundra located in the far north, close to or above the Arctic Circle.

Arid: Dry.

Arroyo: The dry bed of a stream that flows only after rain; also called a wash or a *wadi*.

Artifacts: Objects made by humans, including tools, weapons, jars, and clothing.

Artificial grassland: A grassland created by humans.

Artificial wetland: A wetland created by humans.

Atlantic blanket bogs: Blanket bogs in Ireland that are less than 656 feet (200 meters) above sea level.

Atolls: Ring-shaped reefs formed around a lagoon by tiny animals called corals.

Bactrian camel: The two-humped camel native to central Asia.

Bar: An underwater ridge of sand or gravel formed by tides or currents that extends across the mouth of a bay.

Barchan dunes: Sand dunes formed into crescent shapes with pointed ends created by wind blowing in the direction of their points.

Barrier island: An offshore island running parallel to a coastline that helps shelter the coast from the force of ocean waves.

Barrier reef: A type of reef that lines the edge of a continental shelf and separates it from deep ocean water. A barrier reef may enclose a lagoon and even small islands.

Bathypelagic zone: An oceanic zone based on depth that ranges from 3,300 to 13,000 feet (1,000 to 4,000 meters).

Bathyscaphe: A small, manned, submersible vehicle that accommodates several people and is able to withstand the extreme pressures of the deep ocean.

Bay: An area of the ocean partly enclosed by land; its opening into the ocean is called a mouth.

Beach: An almost level stretch of land along a shoreline.

Bed: The bottom of a river or stream channel.

Benthic: Term used to describe plants or animals that live attached to the seafloor.

Biodiverse: Term used to describe an environment that supports a wide variety of plants and animals.

Bio-indicators: Plants or animals whose health is used to indicate the general health of their environment.

Biological productivity: The growth rate of life forms in a certain period of time.

Biome: A distinct, natural community chiefly distinguished by its plant life and climate.

Blanket bogs: Shallow bogs that spread out like a blanket; they form in areas with relatively high levels of annual rainfall.

Bog: A type of wetland that has wet, spongy, acidic soil called peat.

Boreal forest: A type of coniferous forest found in areas bordering the Arctic tundra. Also called taiga.

Boundary layer: A thin layer of water along the floor of a river channel where friction has stopped the flow completely.

Brackish water: A mixture of freshwater and saltwater.

Braided stream: A stream consisting of a network of interconnecting channels broken by islands or ridges of sediment, primarily mud, sand, or gravel.

Branching network: A network of streams and smaller rivers that feeds a large river.

Breaker: A wave that collapses on a shoreline because the water at the bottom is slowed by friction as it travels along the ocean floor and the top outruns it.

Browsers: Herbivorous animals that eat from trees and shrubs.

Buoyancy: Ability to float.

Buran: Strong, northeasterly wind that blows over the Russian steppes.

Buttresses: Winglike thickenings of the lower trunk that give tall trees extra support.

Canopy: A roof over the forest created by the foliage of the tallest trees.

Canyon: A long, narrow valley between high cliffs that has been formed by the eroding force of a river.

Carbon cycle: Natural cycle in which trees remove excess carbon dioxide from the air and use it during photosynthesis. Carbon is then returned to the soil when trees die and decay.

Carnivore: A meat-eating plant or animal.

Carrion: Decaying flesh of dead animals.

Cay: An island formed from a coral reef.

Channel: The path along which a river or stream flows.

Chemosynthesis: A chemical process by which deep-sea bacteria use organic compounds to obtain food and oxygen.

Chernozim: A type of temperate grassland soil; also called black earth.

Chinook: A warm, dry wind that blows over the Rocky Mountains in North America.

Chitin: A hard chemical substance that forms the outer shell of certain invertebrates.

Chlorophyll: The green pigment in leaves used by plants to turn energy from the sun into food.

Clear-cutting: The cutting down of every tree in a selected area.

Climax forest: A forest in which only one species of tree grows because it has taken over and only that species can survive there.

Climbers: Plants that have roots in the ground but use hooklike tendrils to climb on the trunks and limbs of trees in order to reach the canopy, where there is light.

Cloud forest: A type of rain forest that occurs at elevations over 10,500 feet (3,200 meters) and that is covered by clouds most of the time.

Commensalism: Relationship between organisms in which one reaps a benefit from the other without either harming or helping the other.

Commercial fishing: Fishing done to earn money.

Conifer: A tree that produces seeds inside cones.

Coniferous trees: Trees, such as pines, spruces, and firs, that produce seeds within a cone.

Consumers: Animals in the food web that eat either plants or other animals.

Continental shelf: A flat extension of a continent that tapers gently into the sea.

Continental slope: An extension of a continent beyond the continental shelf that dips steeply into the sea.

Convergent evolution: When distantly related animals in different parts of the world evolve similar characteristics.

Coral reef: A wall formed by the skeletons of tiny animals called corals.

Coriolis effect: An effect on wind and current direction caused by Earth's rotation.

Crustaceans: Invertebrate animals that have hard outer shells.

Current: The steady flow of water in a certain direction.

Dambo: Small marsh found in Africa.

Dark zone: The deepest part of the ocean, where no light reaches.

Deciduous: Term used to describe trees, such as oaks and elms, that lose their leaves during cold or very dry seasons.

Decompose: The breaking down of dead plants and animals in order to release nutrients back into the environment.

Decomposers: Organisms that feed on dead organic materials, releasing nutrients into the environment.

Deforestation: The cutting down of all the trees in a forest.

Dehydration: Excessive loss of water from the body.

Delta: Muddy sediments that have formed a triangular shape over the continental shelf near the mouth of a river.

Deposition: The carrying of sediments by a river from one place to another and depositing them.

Desalination: Removing the salt from seawater.

Desert: A very dry area receiving no more than 10 inches (25 centimeters) of rain during a year and supporting little plant or animal life.

Desertification: The changing of fertile lands into deserts through destruction of vegetation (plant life) or depletion of soil nutrients. Topsoil and groundwater are eventually lost as well.

Desert varnish: A dark sheen on rocks and sand believed to be caused by the chemical reaction between overnight dew and minerals in the soil.

Diatom: A type of phytoplankton with a geometric shape and a hard, glasslike shell.

Dinoflagellate: A type of phytoplankton having two whiplike attachments that whirl in the water.

Discharge: The amount of water that flows out of a river or stream into another river, a lake, or the ocean.

Doldrums: Very light winds near the equator that create little or no movement in the ocean.

Downstream: The direction toward which a river or stream is flowing.

Drainage basin: All the land area that supplies water to a river or stream.

Dromedary: The one-humped, or Arabian, camel.

Drought: A long, extremely dry period.

Dune: A hill or ridge of sand created by the wind.

Duricrusts: Hard, rocklike crusts on ridges that are formed by a chemical reaction caused by the combination of dew and minerals such as limestone.

Ecosystem: A network of organisms that have adapted to a particular environment.

Elfin forest: The upper cloud forest at about 10,000 feet (3,000 meters) which has trees that tend to be smaller, and twisted.

Emergents: The trees that stand taller than surrounding trees.

Epiphytes: Plants that grow on other plants with their roots exposed to the air. Sometimes called "air" plants.

Etermy: A current that moves against the regular current, usually in a circular motion.

Elevation: The height of an object in relation to sea level.

Emergent plants: Plants that are rooted at the bottom of a body of water that have top portions that appear to be above the water's surface.

Emergents: The very tallest trees in the rain forest, which tower above the canopy.

Engineered wood: Manufactured wood products composed of particles of several types of wood mixed with strong glues and preservatives.

Epilimnion: The layer of warm or cold water closest to the surface of a large lake.

Epipelagic zone: An oceanic zone based on depth that reaches down to 650 feet (200 meters).

Epiphytes: Plants that grow on other plants or hang on them for physical support.

Ergs: Arabian word for vast seas of sand dunes, especially those found in the Sahara.

Erosion: Wearing away of the land.

Estivation: An inactive period experienced by some animals during very hot months.

Estuary: The place where a river traveling through lowlands meets the ocean in a semi-enclosed area.

Euphotic zone: The zone in a lake where sunlight can reach.

Eutrophication: Loss of oxygen in a lake or pond because increased plant growth has blocked sunlight.

Fast ice: Ice formed on the surface of the ocean between pack ice and land.

Faults: Breaks in Earth's crust caused by earthquake action.

Fell-fields: Bare rock-covered ground in the alpine tundra.

Fen: A bog that lies at or below sea level and is fed by mineral-rich ground-water.

First-generation stream: The type of stream on which a branching network is based; a stream with few tributaries. Two first-generation streams join to form a second-generation stream and so on.

Fish farms: Farms in which fish are raised for commercial use; also called hatcheries.

Fjords: Long, narrow, deep arms of the ocean that project inland.

Flash flood: A flood caused when a sudden rainstorm fills a dry riverbed to overflowing.

Floating aquatic plant: A plant that floats either partly or completely on top of the water.

Flood: An overflow caused when more water enters a river or stream than its channel can hold at one time.

Floodplain: Low-lying, flat land easily flooded because it is located next to streams and rivers.

Food chain: The transfer of energy from organism to organism when one organism eats another.

Food web: All of the possible feeding relationships that exist in a biome.

Forbs: A category of flowering, broadleaved plants other than grasses that lack woody stems.

Forest: A large number of trees covering not less than 25 percent of the area where the tops of the trees interlock, forming a canopy at maturity.

Fossil fuels: Fuels made from oil and gas that formed over time from sediments made of dead plants and animals.

Fossils: Remains of ancient plants or animals that have turned to stone.

Freshwater lake: A lake that contains relatively pure water and relatively little salt or soda.

Freshwater marsh: A wetland fed by freshwater and characterized by poorly drained soil and plant life dominated by nonwoody plants.

Freshwater swamp: A wetland fed by freshwater and characterized by poorly drained soil and plant life dominated by trees.

Friction: The resistance to motion when one object rubs against another.

Fringing reef: A type of coral reef that develops close to the land; no lagoon separates it from the shore.

Frond: A leaflike organ found on all species of kelp plants.

Fungi: Plantlike organisms that cannot make their own food by means of photosynthesis; instead they grow on decaying organic matter or live as parasites on a host.

Geyser: A spring heated by volcanic action. Some geysers produce enough steam to cause periodic eruptions of water.

Glacial moraine: A pile of rocks and sediments created as a glacier moves across an area.

Global warming: Warming of Earth's climate that may be speeded up by air pollution.

Gorge: A deep, narrow pass between mountains.

Grassland: A biome in which the dominant vegetation is grasses rather than trees or tall shrubs.

Grazers: Herbivorous animals that eat low-growing plants such as grass.

Ground birds: Birds that hunt food and make nests on the ground or close to it.

Groundwater: Freshwater stored in rock layers beneath the ground.

Gulf: A large area of the ocean partly enclosed by land; its opening is called a strait.

Gymnosperms: Trees that produce seeds that are often collected together into cones; most conifers are gymnosperms.

Gyre: A circular or spiral motion.

Hadal zone: An oceanic zone based on depth that reaches from 20,000 to 35,630 feet (6,000 to 10,860 meters).

Hardwoods: Woods usually produced by deciduous trees, such as oaks and elms.

Hatcheries: Farms in which fish are raised for commercial use; also called fish farms.

Headland: An arm of land made from hard rock that juts out into the ocean after softer rock has been eroded away by the force of tides and waves.

Headwaters: The source of a river or stream.

Herbicides: Poisons used to control weeds or any other unwanted plants.

Herbivore: An animal that eats only plant matter.

Lowland rain forest: Rain forest found at elevations up to 3,000 feet (900 meters).

Low tide: A lowering of the surface level of the ocean caused by Earth's rotation and the gravitational pull of the sun and moon.

Macrophytic: Term used to describe a large plant.

Magma: Molten rock from beneath Earth's crust.

Mangrove swamp: A coastal saltwater swamp found in tropical and subtropical areas.

Marine: Having to do with the oceans.

Marsh: A wetland characterized by poorly drained soil and by plant life dominated by nonwoody plants.

Mature stream: A stream with a moderately wide channel and sloping banks.

Meandering stream: A stream that winds snakelike through flat countryside.

Mesopelagic zone: An oceanic zone based on depth that ranges from 650 to 3,300 feet (200 to 1,000 meters).

Mesophytes: Plants that live in soil that is moist but not saturated.

Mesophytic: Term used to describe a forest that grows where only a moderate amount of water is available.

Mid-ocean ridge: A long chain of mountains that lies under the World Ocean.

Migratory: Term used to describe animals that move regularly from one place to another in search of food or to breed.

Mixed-grass prairie: North American grassland with a variety of grass species of medium height.

Montane rain forest: Mountain rain forest found at elevations between 3,000 and 10,500 feet (900 and 3,200 meters).

Mountain blanket bogs: Blanket bogs in Ireland that are more than 656 feet (200 meters) above sea level.

Mouth: The point at which a river or stream empties into another river, a lake, or an ocean.

Muck: A type of gluelike bog soil formed when fully decomposed plants and animals mix with wet sediments.

Muskeg: A type of wetland containing thick layers of decaying plant matter.

Mycorrhiza: A type of fungi that surrounds the roots of conifers, helping them absorb nutrients from the soil.

Neap tides: High tides that are lower and low tides that are higher than normal when the Earth, sun, and moon form a right angle.

Nekton: Animals that can move through the water without the help of currents or wave action.

Neritic zone: That portion of the ocean that lies over the continental shelves.

Nomads: People or animals who have no permanent home but travel within a well-defined territory determined by the season or food supply.

North Atlantic Drift: A warm ocean current off the coast of northern Scandinavia.

Nutrient cycle: Natural cycle in which mineral nutrients are absorbed from the soil by tree roots and returned to the soil when the tree dies and the roots decay.

Oasis: A fertile area in the desert having a water supply that enables trees and other plants to grow there.

Ocean: The large body (or bodies) of saltwater that covers more than 70 percent of Earth's surface.

Oceanography: The exploration and scientific study of the oceans.

Old stream: A stream with a very wide channel and banks that are nearly flat.

Omnivore: Organism that eats both plants and animals.

Ooze: Sediment formed from the dead tissues and waste products of marine plants and animals.

Oxbow lake: A curved lake formed when a river abandons one of its bends.

Rapids: Fast-moving water created when softer rock has been eroded to create many short drops in the channel; also called white water.

Reef: A ridge or wall of rock or coral lying close to the surface of the ocean just off shore.

Rhizomes: Plant stems that spread out underground and grow into a new plant that breaks above the surface of the soil or water.

Rice paddy: A flooded field in which rice is grown.

Riffle: A stretch of rapid, shallow, or choppy water usually caused by an obstruction, such as a large rock.

Rill: Tiny gully caused by flowing water.

Riparian marsh: Marsh usually found along rivers and streams.

Rip currents: Strong, dangerous currents caused when normal currents moving toward shore are deflected away from it through a narrow channel; also called riptides.

River: A natural flow of running water that follows a well-defined, permanent path, usually within a valley.

River system: A river and all its tributaries.

Salinity: The measure of salts in ocean water.

Salt lake: A lake that contains more than 0.1 ounce of salt per quart (3 grams per liter) of water.

Salt pan: The crust of salt left behind when a salt or soda lake or pond dries up.

Saltwater marsh: A wetland fed by saltwater and characterized by poorly drained soil and plant life dominated by nonwoody plants.

Saltwater swamp: A wetland fed by saltwater and characterized by poorly drained soil and plant life dominated by trees.

Saturated: Soaked with water.

Savanna: A grassland found in tropical or subtropical areas, having scattered trees and seasonal rains.

Scavenger: An animal that eats decaying matter.

School: Large gathering of fish.

Sea: A body of saltwater smaller and shallower than an ocean but connected to it by means of a channel; sea is often used interchangeably with ocean.

Seafloor: The ocean basins; the area covered by ocean water.

Sea level: The height of the surface of the sea. It is used as a standard in measuring the heights and depths of other locations such as mountains and oceans.

Seamounts: Isolated volcanoes on the ocean floor that do not break the surface of the ocean.

Seashore: The strip of land along the edge of an ocean.

Secondary succession: Period of plant growth occurring after the land has been stripped of trees.

Sediments: Small, solid particles of rock, minerals, or decaying matter carried by wind or water.

Seiche: A wave that forms during an earthquake or when a persistent wind pushes the water toward the downwind end of a lake.

Seif dunes: Sand dunes that form ridges lying parallel to the wind; also called longitudinal dunes.

Shelf reef: A type of coral reef that forms on a continental shelf having a hard, rocky bottom. A shallow body of water called a lagoon may be located between the reef and the shore.

Shoals: Areas where enough sediments have accumulated in the river channel that the water is very shallow and dangerous for navigation.

Shortgrass prairie: North American grassland on which short grasses grow.

Smokers: Jets of hot water expelled from clefts in volcanic rock in the deep-seafloor.

Soda lake: A lake that contains more than 0.1 ounce of soda per quart (3 grams per liter) of water.

Softwoods: Woods usually produced by coniferous trees.

Sonar: The use of sound waves to detect objects.

Soredia: Algae cells with a few strands of fungus around them.

Source: The origin of a stream or river.

Spit: A long narrow point of deposited sand, mud, or gravel that extends into the water.

Spores: Single plant cells that have the ability to grow into a new organism.

Sport fishing: Fishing done for recreation.

Spring tides: High tides that are higher and low tides that are lower than normal because the Earth, sun, and moon are in line with one another.

Stagnant: Term used to describe water that is unmoving and contains little oxygen.

Star-shaped dunes: Dunes created when the wind comes from many directions; also called stellar dunes.

Steppe: A temperate grassland found mostly in southeast Europe and Asia.

Stone circles: Piles of rocks moved into a circular pattern by the expansion of freezing water.

Straight stream: A stream that flows in a straight line.

Strait: The shallow, narrow channel that connects a smaller body of water to an ocean.

Stream: A natural flow of running water that follows a temporary path that is not necessarily within a valley; also called brook or creek. Scientists often use the term to mean any natural flow of water, including rivers.

Subalpine forest: Mountain forest that begins below the snow line.

Subduction zone: Area where pressure forces the seafloor down and under the continental margin, often causing the formation of a deep ocean trench.

Sublittoral zone: The seashore's lower zone, which is underwater at all times, even during low tide.

Submergent plant: A plant that grows entirely beneath the water.

Subpolar gyre: The system of currents resulting from winds occurring near the poles of Earth.

Subsistence fishing: Fishing done to obtain food for a family or a community.

Subtropical: Term used to describe areas bordering the equator in which the weather is usually warm.

Subtropical gyre: The system of currents resulting from winds occurring in subtropical areas.

Succession: The process by which one type of plant or tree is gradually replaced by others.

Succulents: Plants that appear thick and fleshy because of stored water.

Sunlit zone: The uppermost part of the ocean that is exposed to light; it reaches down to about 650 feet (200 meters) deep.

Supralittoral zone: The seashore's upper zone, which is never underwater, although it may be frequently sprayed by breaking waves; also called the splash zone.

Swamp: A wetland characterized by poorly drained soil, stagnant water, and plant life dominated by trees.

Sward: Fine grasses that cover the soil.

Swell: Surface waves that have traveled for long distances and become more regular in appearance and direction.

Taiga: Coniferous forest found in areas bordering the Arctic tundra; also called boreal forest.

Tallgrass prairie: North American grassland on which only tall grass species grow.

Tannins: Chemical substances found in the bark, roots, seeds, and leaves of many plants and used to soften leather.

Tectonic action: Movement of Earth's crust, as during an earthquake.

Temperate bog: Peatland found in temperate climates.

Temperate climate: Climate in which summers are hot and winters are cold, but temperatures are seldom extreme.

Temperate zone: Areas in which summers are hot and winters are cold but temperatures are seldom extreme.

Thermal pollution: Pollution created when heated water is dumped into the ocean. As a result, animals and plants that require cool water are killed.

Thermocline: Area of the ocean's water column, beginning at about 1,000 feet (300 meters), in which the temperature changes very slowly.

Thermokarst: Shallow lakes in the Arctic tundra formed by melting permafrost; also called thaw lakes.

Tidal bore: A surge of ocean water caused when ridges of sand direct the ocean's flow into a narrow river channel, sometimes as a single wave.

Tidepools: Pools of water that form on a rocky shoreline during high tide and that remain after the tide has receded.

Tides: Rhythmic movements, caused by Earth's rotation and the gravitational pull of the sun and moon, that raise or lower the surface level of the oceans.

Tombolo: A bar of sand that has formed between the beach and an island, linking them together.

Trade winds: Winds occurring both north and south of the equator to about 30 degrees latitude; they blow primarily east.

Transverse dunes: Sand dunes lying at right angles to the direction of the wind.

Tree: A large woody perennial plant with a single stem, or trunk, and many branches.

Tree line: The elevation above which trees cannot grow.

Tributary: A river or stream that flows into another river or stream.

Tropical: Term used to describe areas close to the equator in which the weather is always warm.

Tropical tree bog: Bog found in tropical climates in which peat is formed from decaying trees.

Tropic of Cancer: A line of latitude about 25 degrees north of the equator.

Tropic of Capricorn: A line of latitude about 25 degrees south of the equator.

Tsunami: A huge wave or upwelling of water caused by undersea earthquakes that grows to great heights as it approaches shore.

Tundra: A cold, dry, windy region where trees cannot grow.

Turbidity current: A strong downward-moving current along the continental margin caused by earthquakes or the settling of sediments.

Turbine: An energy-producing engine.

Tussocks: Small clumps of vegetation found in marshy tundra areas.

Typhoon: A violent tropical storm that begins over the ocean.

Understory: A layer of shorter, shade-tolerant trees that grow under the forest canopy.

Upstream: The direction from which a river or stream is flowing.

Upwelling: The rising of water or molten rock from one level to another.

Veld: Temperate grassland in South Africa.

Venom: Poison produced by animals such as certain snakes and spiders.

Vertebrates: Animals with a backbone.

Wadi: The dry bed of a stream that flows only after rain; also called a wash or an *arroyo*.

Warm-bodied fish: Fish that can maintain a certain body temperature by means of a special circulatory system.

Wash: The dry bed of stream that flows only after rain; also called an *arroyo* or a *wadi*.

Water column: All of the waters of the ocean, exclusive of the sea bed or other landforms.

Water cycle: Natural cycle in which trees help prevent water runoff, absorb water through their roots, and release moisture into the atmosphere through their leaves.

Waterfall: A cascade of water created when a river or stream falls over a cliff or erodes its channel to such an extent that a steep drop occurs.

Water table: The level of groundwater.

Waves: Rhythmic rising and falling movements in the water.

Westerlies: Winds occurring between 30 degrees and 60 degrees latitude; they blow in a westerly direction.

Wet/dry cycle: A period during which wetland soil is wet or flooded followed by a period during which the soil is dry.

Wetlands: Areas that are covered or soaked by ground or surface water often enough and long enough to support plants adapted for life under those conditions.

Wet meadows: Freshwater marshes that frequently dry up.

Xeriscaping: Landscaping method that uses drought tolerant plants and efficient watering techniques.

Xerophytes: Plants adapted to life in dry habitats or in areas like salt marshes or bogs.

Young stream: A stream close to its headwaters that has a narrow channel with steep banks.

Zooplankton: Animals, such as jellyfish, corals, and sea anemones, that float freely in ocean water.

River and Stream

A river is a natural flow of running water that follows a well-defined, permanent path, usually within a valley. A stream (also called a brook or a creek) is a natural flow of water that follows a more temporary path that is usually not in a valley. The term stream is often used to mean any natural flow of water, including rivers. Although some rivers are larger than some streams, size is not a distinguishing factor.

The origin of a river or stream is called its source. If its source consists of many smaller streams coming from the same region, they are called headwaters. Its channel is the path along which it flows, and its banks are its boundaries, the sloping land along each edge between which the water flows. The point where a stream or river empties into a lake, a larger river, or an ocean, is its mouth. When one stream or river flows into another, usually larger, stream or river, and adds its flow, it is considered a tributary of the larger river. Many tributaries make up a river system. The vast Amazon system in South America, for example, is fed by at least 1,000 tributaries.

The world's longest river is the Nile in eastern Africa. It begins in Ethiopia and travels 4,000 miles (6,437 kilometers) into Egypt, where it discharges its waters into the Mediterranean Sea. The river system that obtains water from the largest area, 2,722,000 square miles (7,077,200 kilometers), is that of the Amazon. The Amazon River transports the largest volume (20 percent) of all the water in all the rivers of the world.

How Rivers and Streams Develop

Rivers and streams are part of Earth's hydrologic cycle. The hydrologic cycle describes the manner in which molecules of water evaporate, condense and form clouds, and return to Earth as precipitation (rain, sleet, or snow). Rivers pass through several stages of development.

WORDS TO KNOW

Arroyo: The dry bed of a stream that flows only after rain; also called a wash or a *wadi*.

Erosion: Wearing away of the land.

Headwaters: The source of a river or stream.

Oxbow lake: A curved lake formed when a river abandons one of its bends.

Rhizomes: Plant stems that spread out underground and grow into a new plant that breaks above the surface of the soil or water.

Riffle: A stretch of rapid, shallow, or choppy water usually caused by an obstruction, such as a large rock.

Rill: Tiny gully caused by flowing water.

Tributary: A river or stream that flows into another river or stream.

Wadi: The dry bed of a stream that flows only after rain; also called a wash or an *arroyo*.

Formation   Rivers and streams owe their existence to precipitation, lakes, and groundwater, combined with gravity and a sloping terrain.

When rain falls on the land, often the soil cannot absorb it all. Much of rain runs off and travels downhill with the aid of gravity, creating rills (tiny gullies). Many of these rills may meet at some point and their waters run together to form bigger gullies until all this water reaches a valley or gouges out its own large channel. When enough water is available to maintain a steady ongoing flow, a stream or river results. Gravity and the pressure of the flowing water cause the river to travel until it is either blocked, in which case the water backs up and forms a lake, or empties into an existing lake or ocean.

Most of the precipitation that feeds streams and rivers comes from runoff. Precipitation may also be stored as ice in glaciers in arctic regions or on mountaintops. As the glaciers melt, they nourish streams, and the streams feed rivers. The Rhine River in Germany, for example, obtains much of its water from the Rheinwaldhorn Glacier in the Swiss Alps.

A lake can be a source of river water. If the land slopes away from the lake at some point and the water level is high enough for it to overflow, a river or stream may form.

Another source of river water is groundwater. Groundwater is water that has seeped beneath Earth's surface where it becomes trapped in layers of rock called aquifers. The Ogallala Aquifer, the largest aquifer in North America, under the Great Plains in the United States is an example. When

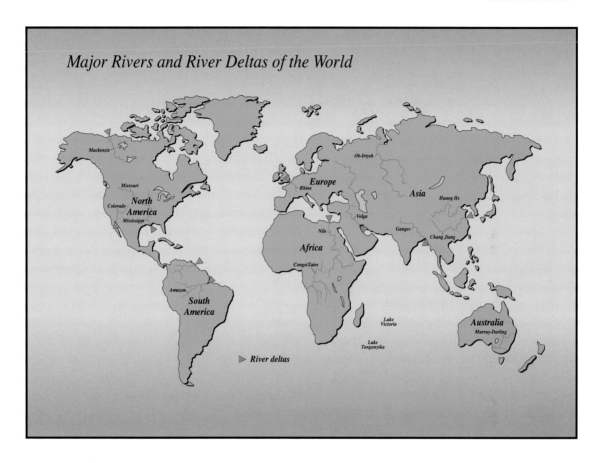

Major Rivers and River Deltas of the World

an aquifer is full, its water escapes to the surface, either by seeping directly into a river or stream bed or by forming a spring (an outpouring of water), which may then become a river's source. As much as 30 percent of the world's freshwater comes from groundwater. It is estimated that groundwater supplies about half of the water in the Mississippi River. In contrast, river water may seep into the ground and fill an aquifer, as the Colorado River does as it travels through Arizona, Nevada, and California.

Stages of development Rivers are said to age not in terms of how long they have existed but in terms of their development as they travel across the land. They are young, mature, or old. Some rivers, such as the Mississippi, may be in all three stages of development at one time.

Young rivers usually occur in highland or mountainous regions and have narrow, rocky channels with many boulders. The water may form

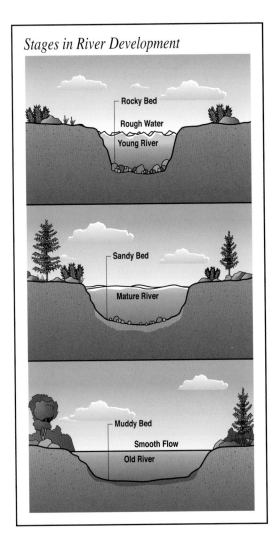

Stages in River Development

Rocky Bed
Rough Water
Young River

Sandy Bed
Mature River

Muddy Bed
Smooth Flow
Old River

waterfalls or foam and gurgle as it rushes over the rocks. Young rivers have few tributaries.

Mature rivers are those that have reached flat land as their tributaries pour more water into them. They often become flooded during periods of heavy rain or snowmelt. Their beds tend to be muddy rather than rocky because of the sediment (particles of sand or soil) carried into them by swift-flowing streams, and the valleys through which they flow are usually fairly broad. Few waterfalls occur on mature rivers, but they may have many bends or loops as they curl across the land looking for the lowest level to follow.

A river is old near its mouth, where layers of sediment build up over time. Here, the land is wide and flat, and the water travels more smoothly than in young and mature rivers.

Kinds of Rivers and Streams

Rivers and streams can be classified according to their degree of permanence, the shape of their channels, and their branching network.

Degree of permanence Permanent streams and rivers flow all year long. Enough water is available to keep them from drying up completely, even during a long, dry spell. Most large rivers are permanent.

Intermittent streams and rivers are seasonal. They occur only during the rainy season or in the spring after the snow melts. Rivers and streams in desert regions tend to be intermittent, where they are also called wadis or arroyos.

Interrupted streams and rivers flow above ground in some places and then disappear from sight as they dip down under sand and gravel to flow underground in other places. The Santa Fe River in Florida is an example of an interrupted river.

During the wet season, this dry river bed could be a flowing stream. IMAGE COPYRIGHT LINDA ARMSTRONG, 2007. USED UNDER LICENSE FROM SHUTTERSTOCK.COM.

Channel shape   The material over which a stream or river flows, and the force of the water as it travels determine the shape of the channel, which can be straight, braided, or meandering.

Straight streams or rivers flow in a straight line. This type is very rare because flowing water tends to trace a weaving path. Straight streams that do occur tend to have rocky channels.

Braided streams or rivers, such as the Platte River in Nebraska, consist of a network of interconnecting channels broken by islands or ridges of sediment, primarily mud, sand, or gravel. Braided streams often occur in highland regions and have a steep slope.

Meandering streams or rivers wind snakelike through relatively flat countryside. Meandering streams tend to have low slopes and soft channels of silt (soil) or clay. The Menderes River in Turkey meanders, and the term, which means to wander aimlessly, is derived from its name.

Branching network   Rivers do not usually originate from a single spot but are the result of a branching network that resembles the branches on a tree. The smallest streams or branches that do not have tributaries are called first-order streams. If two first-order streams join to form a new stream, the new stream is considered a second-order stream. When two second-order streams join, they form a third-order stream, and so on. This system of ordering was developed by an American engineer, Robert E. Horton (1875–1945) in 1945. A very large river the size of the

Architects of the Underground

In areas where large deposits of limestone and other soft rock are found, underground rivers may create vast networks of caves. The best example is Mammoth Cave in Kentucky, which has large chambers and underground passageways on five levels that wind back and forth for a total length of 365 miles (587 kilometers). Several underground rivers and streams still flow through the cave, including Echo River, on which tour boats were once operated. The beautiful caverns and rock formations carved by the rivers have been given intriguing names, such as King Solomon's Temple, the Pillars of Hercules, and the Giant's Chamber. Ancient Native American tribes once used the caves, and their mummified bodies have been found there.

Mississippi is usually considered a tenth-order stream. The Amazon River is rated a twelfth- or thirteenth-order stream.

The Water Column

The water column refers to the water in a river or stream, exclusive of its channel (path) or banks. All of the world's rivers and streams combined contain less than 1 percent of all the water on Earth. Most water is held in the oceans.

Composition Rivers and streams carry fresh water, but this water may be clear or cloudy, polluted or clean. Its characteristics are determined by where it originates and the nature of its channel.

As the water moves across the land, particles of rock, soil, and decaying plant or animal matter, called sediment, become suspended in it. If a river carries a large quantity of sediment, it will be muddy. The smaller the amount of sediment, the clearer the water.

River water reacts chemically with the rocks in its channel and dissolves some of the minerals they contain, called salts. These dissolved

This is the braided Rakaia River in Canterbury, New Zealand. IMAGE COPYRIGHT CHRIS HELLYAR, 2007. USED UNDER LICENSE FROM SHUTTERSTOCK.COM.

salts give the water its taste and make it "hard." Rivers and streams carry large amounts of these salts. The Niagara River, which lies on the border between the state of New York and Ontario, Canada, carries an average of 60 tons (54 metric tons) of dissolved mineral salts per minute as it pours over Niagara Falls, giving the water a green color.

Circulation   Water in a river or stream is almost always in motion. The direction from which it comes is called upstream, and the direction toward which it flows is called downstream. If you paddle upstream, you ascend the river. If you paddle downstream, you descend.

Velocity   Rivers and streams acquire their initial energy from their elevation. Their waters are always falling downhill toward their ultimate goal, sea level (the average level of the surface of the sea). The fastest velocities (speeds) occur at waterfalls. The water flowing over Niagara Falls has been clocked at 68 miles (109 kilometers) per hour. Most rivers and streams seldom exceed 10 miles (1.6 kilometers) per hour, and the average speed is 2 to 4 miles (3.2 to 6.4 kilometers) per hour.

Although gravity helps get things started, gushing mountain streams actually move more slowly than large, broad, mature rivers. This happens because friction (resistance to motion when one object rubs against another) caused by water molecules rubbing against the channel and banks decreases downstream where the channel is usually smoother and less rocky. Where the river is wider, deeper, and the volume of water is greater, a smaller percentage of water molecules is exposed to friction.

The water at different locations or levels in the same section of river may move at varying speeds. The velocity is fastest in the middle and just below the surface. Movement slows down with depth and along the banks because friction increases. Along the very bottom of the channel is the boundary layer, a layer of water only 0.1 inch (0.25 centimeter) deep, where friction has stopped the flow completely.

Wave and current   Waves are rhythmic rising and falling movements in the water. Surface waves are caused by wind blowing across a river's surface. Winds tend to follow a regular pattern. They occur

Mark Twain

Samuel Clemens (1835–1910), a famous American author, wrote under the name Mark Twain. The word twain means "two fathoms deep" (a fathom is about 6 feet [2 meters]). It was one of the terms shouted by crewmen on Mississippi river boats who were measuring the river's depth to help the captain know if the boat could pass through that part of the channel. Clemens himself worked for a time as a steamboat pilot, and many of his stories have to do with the river. His river adventures are chronicled in *Life on the Mississippi* and *The Adventures of Huckleberry Finn*.

This high altitude lake in the Misty Fjords National Monument and Wilderness in Ketchikan, Alaska, gives the water energy as it cascades into lower terrain. COPYRIGHT © 2003 KELLY A. QUIN.

in the same place and blow in the same direction, and the movement of waves follows this pattern.

The current is the steady flow of the water in a particular direction, usually from upstream to downstream. River currents can be influenced by the slope and composition of the channel and the position of landmasses. When a current travels down a very steep slope, it may gain speed and force. When it meets a landmass, such as an island, it may be deflected and turn in a new direction.

An eddy is a current that moves against the regular current, usually in a circular, or whirlpool, motion. A riffle is a stretch of rapid, shallow, or choppy water that alternates with a quiet pool. A riffle is usually caused as the current flows over stones or gravel. A quiet pool is a deep, still area of water.

Discharge   A river's discharge is the amount of water that flows out of it over a given period of time into another river, a lake, or the ocean. All of the world's rivers and streams combined discharge a total of 28,000,000,000 gallons (106,000,000,000 liters) of water every day. If all this water were to flow out over the land, it would cover the land to a depth of 9 inches (25 centimeters).

Discharge is usually measured by multiplying the river's width at the surface by its average depth times its velocity. The Amazon, which empties into the Atlantic Ocean, discharges more water than any other river—more than 52,834,410 gallons per second (199,999,999 liters per second).

Tidal bores   Tidal bores are surges of ocean water caused when ridges of sand direct the ocean's flow back into a narrow river channel, sometimes as a single wave. Most tidal bores are harmless. The Tsientang River in China, however, has bore speeds of 18 to 25 miles (29 to 40 kilometers) per hour.

Floods   Floods are caused when more water enters the river than its channel can hold. The result is increased discharge and high water levels.

Most rivers overflow their banks every two to three years, but the amount of overflow is usually moderate. Depending on the river's origins and location, flooding is caused by melting snow, melting glaciers, or heavy rainfall, and sometimes by all three. Rivers supplied by snow or glaciers ordinarily produce only moderate flooding because snow and glaciers melt slowly and the extra water does not enter the river all at once. When heavy rainfall is added the discharge may be enormous. Rain in a location upstream may cause a flood downstream in a region that received no rain at all because rivers are so long.

The seriousness of a flood is measured by comparing the river's average annual discharge rate to the rate during the flood. The Huang He (Yellow) River in China, for example, has a discharge rate of 88,286 to 158,916 cubic feet (2,500 to 4,500 cubic meters) per second. During a flood in 1958, its rate increased to 812,237 cubic feet (23,000 cubic meters). The flood covered 34,000 square miles (88,060 square kilometers) and killed 1 million people. The worst flood to occur in the United States was in 1993,

Yukon Gold Rush

Between 1860 and 1910, people stampeded to the Yukon Territory of Canada in search of gold. They found it not only in the ground, but also in the sediments of rivers, especially the Klondike, the Lewes, the Stewart, and the lower Yukon. These sediments were scooped up into a flat pan and sloshed with water in an effort to separate bits of rock and other debris from the shining gold particles. It is estimated that about $100 million in gold was taken from the Yukon Territory during that time.

Most rivers overflow their banks every few years. Here is a flooded parkland at Henley on Thames during the UK floods of summer 2007. IMAGE COPYRIGHT PETER ELVIDGE, 2007. USED UNDER LICENSE FROM SHUTTERSTOCK.COM.

Amazon River Name Origin

The first European to explore the Amazon and its basin was Francisco de Orellana of Spain (c. 1511–1546). In 1541, he made a seventeen-month trip from the foot of the Andes Mountains in the west to the mouth of the Amazon River in the east. The journey was difficult and involved many battles with the native peoples. At times food was scarce and he and his men had to eat toads and snakes. This lack of food may partly explain why de Orellana thought they had been attacked by a band of female warriors resembling the famous Amazons of Greek legends. Other explorers failed to encounter these hostile females, but de Orellana had named the river in their honor, and the name stuck.

along the middle and lower portions of the Mississippi River. Killing 487 people, the flood caused $15 billion in property damage.

In very dry regions, a river or stream may dry up completely for several weeks or months. During a sudden rainstorm, the channel may not be large enough to contain the amount of water, and a flash flood may occur. Flash floods are dangerous because they happen very suddenly. This wall of water brings down everything it reaches.

Effect on climate and atmosphere The climate of a stream or river depends upon its location as it travels over land. In general, if it passes through a desert area, such as Northern Africa, the climate will be hot and dry. If it begins on top of a mountain in the Northern Hemisphere, the climate will be cold. Rivers and streams in temperate (moderate) climates are often affected by seasonal changes. In these areas, small streams or rivers may dry up in summer or develop a layer of ice in winter.

The presence of a large river can create some climatic differences. In summer, evaporation may create more moisture in the air. When the water temperature is cooler than the air temperature, winds off the river can help cool the nearby region. The water in rivers and streams is partly responsible for the precipitation that falls on land. The water evaporates in the heat of the sun, forms clouds, and falls elsewhere.

Geography of Rivers and Streams

Rivers and streams are strong forces in shaping the landscape through which they flow. As the current moves against the channel and banks, water and the particles of sediment the river carries wear away the surface with a cutting action called erosion (ee-ROH-zuhn).

The faster a river flows, the faster it wears the land away and the more sediment it bears. Some of the eroded chunks and particles may sink to the bottom. Others are carried along and, as the river slows down, are dropped farther downstream. Erosion (wearing away) and deposition (dep-oh-ZIH-shun; layering) make many changes over time.

Waterfalls, such as these on the Tahquamenon River in Michigan, occur when a river or stream falls over a cliff or erodes the channel to such an extent that a steep drop occurs. COPYRIGHT © 2006 KELLY A. QUIN.

The drainage basin A river or stream obtains runoff from an area of land called its drainage basin, or watershed. The drainage basin can be extremely large. The Mississippi-Missouri river system, for example, has as its drainage basin the entire area between the Appalachian Mountains in the east and the Rocky Mountains in the west. The world's largest drainage basin is that of the Amazon, which measures 2,700,000 square miles (7,000,000 square kilometers).

Channel and banks The channel of a river or stream slopes down to the center where it is deepest. The bottom of the channel is called its bed. Channel beds may be rocky or covered with gravel, pebbles, or mud. Some river bed sediments are formed from the waste products and dead tissues of animals and plants. Others usually consist of clay, stone, and other minerals.

The general shape of the banks is partly determined by the general shape of the surrounding land. If the river runs through a plain, then its banks will be broad and level. If it occurs in the mountains, then its banks may be steep and rocky.

As water rushes through the channel of a young river, erosion tends to scour it out and make it deeper. Young river valleys are often narrow and steep sided. As the river matures and velocity and volume increase, width tends to increase more than depth. A mature valley tends to have a gentle slope, and an old valley may be almost flat.

MAJOR RIVER SYSTEMS

Name	Location	Length (mi./km.)	Basin Area (sq. mi./sq. km.)
Nile	North Africa	4,157 / 6,651	1,100,000/2,860,000
Amazon	South America	3,915 / 6,437	2,722,000/7,077,200
Mississippi-Missouri	North America	3,860 / 6,176	1,243,700/3,233,620
Ob-Irtysh	Siberia	3,461 / 5,538	959,500/1,919,000
Yangtze	China	3,434 / 5,494	756,500/1,919,000
Murray-Darling	Australia	3,371 / 5,394	414,253/1,077,058
Yenisei-Angara	Siberia	3,100 / 4,960	1,003,000/2,607,800
Zaire-Congo	Africa	2,716 / 4,346	1,425,000/3,705,000

Moderate periodic flooding is a major cause of changes in bank shape. Although large floods cause more erosion, they do not occur often enough to make drastic or permanent changes.

Landforms   Landforms created by rivers and streams include floodplains and levees; wetlands; canyons and gorges; rapids and waterfalls; spits, bars, and shoals; and deltas and estuaries.

Floodplains and levees   When a stream carrying a large quantity of sediment flows into another stream or out over a plain, it slows down and its sediments settle to the bottom. These sediments (called alluvium [a-LOO-VEE-um]) often spread out in a fan shape, creating an alluvial fan. When this occurs frequently, the result may be a floodplain. A floodplain is a flat region with soil enriched by river sediments and usually makes good farmland.

Very silty rivers may create high banks of sediments called levees. The levees of the Mississippi River are several yards (meters) high in places.

Wetlands   Sometimes a river may change its path during a flood. The current may seek out a new path or, as the flood waters recede, sediments prevent the water in a loop or bend from rejoining the rest of the river. Here, a wetland, such as a marsh or swamp, or an oxbow lake may form. Oxbow lakes have a curved shape, which gives a clue to the origin of their name. (An oxbow is a U-shaped piece that loops around an ox's neck, and is part of the harness.)

Canyons and gorges   Many riverbanks consist of both hard and soft rock. Moving water erodes the soft rock first, sometimes sculpturing strange shapes. In regions where much of the rock is soft, rivers and streams may cut deep canyons (long, narrow valleys between high cliffs) and gorges (deep, narrow passes). The Gooseneck Canyon in Utah was cut into the soft limestone of the region by the San Juan River over millions of years. The distance from the top of the canyon to the river is now 1,500 feet (456 meters).

Caves may be gradually carved into the sides of cliffs by erosion. In large rivers, headlands may be created. A headland is an arm of land made of hard rock that juts out from the bank into the water after softer rock has been eroded away.

Rapids and waterfalls   When fast-moving water erodes softer rock downstream, gradually cutting away the bed in some places and creating many short drops, rapids may form. Rapids are often called "white water" because of the foam created when the rushing water hits the exposed bands of rock. The Salmon River in Idaho flows through steep canyons and has many rapids. Its rapids are so dangerous to travelers that it has been nicknamed the "River of No Return."

This magnificent waterfall is in the Yellowstone National Park. IMAGE COPYRIGHT GSK, 2007. USED UNDER LICENSE FROM SHUTTERSTOCK.COM.

When a river or stream falls over a cliff or erodes the channel to such an extent that a steep drop occurs, it creates a waterfall. The world's highest waterfall is Angel Falls on the Rio Churun in southeastern Venezuela. Its greatest drop is 3,212 feet (979 meters). Victoria Falls, on the Zambezi River between Zimbabwe and Zambia, is considered one of the Seven Natural Wonders of the World. It is the world's largest falls at nearly 4,400 feet (5,577 meters) wide. Africans call it *Mosi-oa-tunya*, the "smoke that thunders."

Where a waterfall strikes the valley below, it gouges out a deep basin called the plunge pool. Eventually, it erodes the lip of rock over which it flows. As erosion continues over time, the waterfall slowly moves upstream. Niagara Falls, for example, is being cut back at a rate of 3 feet (1 kilometer) a year.

Spits, bars, and shoals   A spit is a long, narrow point of deposited sand, mud, or gravel that extends into the water. A bar is an underwater

Fast moving water erodes softer rock downstream, creating many short drops, called rapids. IMAGE COPYRIGHT ANDREW MCDONOUGH, 2007. USED UNDER LICENSE FROM SHUTTERSTOCK.COM.

ridge of sand or gravel, formed by currents, that extends across a channel or the inside bank of a curve. Shoals are areas where enough sediments have accumulated that the water is very shallow and dangerous for navigation.

Deltas and estuaries   Where rivers meet a lake or ocean, huge amounts of silt can be deposited along the shoreline. Large rivers can dump so much silt that islands of mud build up, eventually forming a triangle-shaped area called a delta. The finer debris that does not settle as quickly may drift around, making the water cloudy. The Nile River in Egypt and the Ganges (GANN-jeez) in India both have large deltas. The Ganges Delta is 220 miles (350 kilometers) wide and covers 25 percent of India's territory, making it the largest delta in the world.

When a river traveling through lowlands meets the ocean in a semi-enclosed channel or bay, the area is called an estuary. The water in an estuary is brackish—a mixture of fresh and salt water. In these gently sloping areas, river sediments collect and muddy shores form. Chesapeake

A spit is a long narrow point of deposited sand, mud, or gravel that extends into the water.
COPYRIGHT © 2006 KELLY A. QUIN.

Bay, along the Atlantic Coast of Maryland, is the largest estuary in the United States. Estuaries are often sites for harbors because they can serve both river and ocean travel.

Elevation   Rivers and streams occur at all altitudes. Their altitude changes over their length because they usually begin in highlands or mountains and move to lower elevations. The headwaters of the Ganges River in India, for example, start from an ice cave in the Himalaya Mountains at an elevation of 10,300 feet (16,480 meters). A thousand miles (1,600 kilometers) from its mouth, the Ganges is only about 600 feet (180 meters) above sea level, and 0 feet (0 meters) at its mouth where it enters the Bay of Bengal.

Plant Life

Few plants can root and grow in running water. Most plants that live in rivers and streams are found along the banks or in quiet pools where the environment is similar to that of a lake or pond. Plants with floating leaves do not do well in fast-moving streams because the current pulls the leaves under. When rooted plants gain a foothold and manage to reproduce, they can grow so numerous that they slow the flow of water and can even cause flooding.

River and stream plants may be classified as submergent, floating aquatic (water), or emergent according to their relationship with the water.

A submergent plant grows beneath the water; even its leaves lie below the surface. Submergents include the New Zealand pygmyweed, tape grass, and water violet.

Floating aquatics float on the water's surface. Some, such as duckweed, have no roots. Others, such as the water lily, have leaves that float on the surface, stems that are underwater, and roots that are anchored to the bottom.

An emergent plant grows partly in and partly out of the water. The roots are usually underwater, but the stems and leaves are at least partially exposed to air. They have narrow, broad leaves, and some even produce flowers. Emergents include reeds, rushes, grasses, cattails, and sedges.

Plants that live in rivers and streams can be divided into four main groups: bacteria; algae (AL-jee); fungi (FUHN-ji) and lichens (LY-kens); and green plants.

Algae   Most submergent plants are algae. (It is generally recognized that not all algae fit neatly into the plant category.) Some forms of algae are so tiny they cannot be seen without the help of a microscope. These, like spirogyra, sometimes float on the water in slow-moving rivers or coat the surface of rocks, other plants, and river debris. Other species, such as blanket weed, are larger and remain anchored to the riverbed in quiet pools.

Most algae have the ability to make their own food by photosynthesis (foh-toh-SIN-thih-sihs), the process by which plants use energy from sunlight to change water and carbon dioxide from the air into the sugars and starches they use for food. A by-product of photosynthesis is oxygen, which combines with the water and enables aquatic animals, such as fish, to breathe. Algae require other nutrients that must be found in the water, such as nitrogen, phosphorus, and silicon. Algal growth increases when nitrogen and phosphorus are added to a stream by sewage or by runoff from fertilized farmland.

Common river and stream algae   Phytoplankton float on the surface of the water. Two types of phytoplankton, diatoms and dinoflagellates (dee-noh-FLAJ-uh-lates), are the most common. Diatoms have simple, geometric shapes and hard, glasslike cell walls. They live in colder regions and even within arctic ice. Dinoflagellates have two whiplike attachments that make a swirling motion. They live in tropical regions (regions around the equator).

Macrophytic algae are not as common in fast-moving water, but some filamentous species can be found growing on stable surfaces.

Growing season   Algae contain chlorophyll, a green pigment used during photosynthesis. As long as light is available, algae can grow. Growth is often seasonal. In some areas, such as the Northern Hemisphere, most growth occurs during the summer months when the sun is more directly overhead. In temperate zones, growth peaks in the spring but continues throughout the summer. In regions near the equator, no growth peaks occur since it is warm year round. As a result, growth is steady throughout the year.

Reproduction   Algae may reproduce in one of three ways. Some split into two or more parts, each part becoming a new, separate plant. Others form spores (single cells that have the ability to grow into a new organism), and a few reproduce sexually, during which cells from two different plants unite to create a new plant.

Fungi and lichens

As with algae, it is generally recognized that fungi and lichens do not fit neatly into the plant category. Fungi are plantlike organisms that cannot make their own food by means of photosynthesis. Instead, they grow on decaying organic (material derived from living organisms) matter or live as parasites (organisms that depend upon other organisms for food or other needs) on a host. Fungi grow best in a damp environment and are often found along riverbanks. Common fungi include mushrooms, rusts, and puffballs.

Lichens are combinations of algae and fungi that tend to grow on rocks and other smooth surfaces. The alga produces the food for both itself and the fungus by means of photosynthesis. The fungi may provide moisture for the algae. One of the most common lichens found near streams is called reindeer moss, which prefers the solid surfaces of rocks.

Green plants

Most green plants need several basic things to grow: light, air, water, warmth, and nutrients. Near a stream, light and water are in plentiful supply. Nutrients, primarily nitrogen, phosphorus, and potassium, are usually obtained from the soil. Some soils are lower in these nutrients. They may also be low in oxygen. These deficiencies may help limit the kinds of plants that can grow in an area.

The true green plants found growing in the water along the banks of rivers and streams are similar to those that grow on dry land. Unlike algae, they have roots and some, like the eelgrass, may even bloom underwater.

Large beds of green plants slow the movement of water and help prevent erosion. Some water animals use green plants for food and for hiding places. Plants that occupy the emergent zone between the water-covered area and dry land, such as sallows (European species of willows) and sedges, need moist, but not saturated, soil.

Common river and stream green plants   Common plants found in and around rivers and streams include willow moss, water lily, pondweed, duckweed, aquatic buttercup, great pond sedge, water plantain, water lettuce, mosquito fern, water soldier, bur reed, marsh horsetail, reed mace, mare's tail, greater spearwort, flowering rush, water forget-me-not, St. John's wort, alder, and weeping willow.

Growing season   Climate and the amount of precipitation both affect the length of the growing season. Warmer temperatures and moisture usually signify the beginning of growth. In regions that are colder or receive little rainfall, the growing season is short. Growing conditions are also affected by the amount of moisture in the soil, which can range from saturated, during flooding or a rainy season, to dry.

Reproduction   Green plants reproduce by several methods. One is pollination, in which the pollen from the male reproductive part (called a stamen in flowering plants) of a plant is carried by wind or insects to the female reproductive part (called a pistil in flowering plants). Water lilies, for example, are closed in the morning and evening, but open during midday when the weather is warmer and insects are more active and more likely to visit. Some shoreline plants, like sweet flag, reproduce by sending out rhizomes (RY-zohms), which are stems that spread out under water or soil and form new plants. Reed mace and common reeds reproduce by this means.

Endangered species   Changes in the habitat, such as pollution and the presence of dams, endanger river and stream plants. Plants are also endangered by people who collect them, and by the use of fertilizers and herbicides (poisons used to control weeds), which enter the water as runoff from farms.

The papyrus plant in Egypt, for example, is endangered because of the Aswan High Dam. The dam traps Nile water in its reservoir upstream, and the swamps and ponds that form the habitat for papyrus are disappearing.

Animal Life

In addition to the land animals that visit for food and water, rivers and streams support many species of aquatic animals. Some swim freely in the water, while others live along the muddy bottom. Some prefer life in midstream or on rocks beneath a fast-moving flow; others seek the shallows or quiet pools. Shallows are often warm and exposed to sunlight, but animals that prefer cool shady spots can find them along the banks beneath overhanging trees. Many river and stream animals are adapted to life under high-speed conditions. Salmon and trout, for example, are torpedo-shaped and can swim against the current more easily than other fish. Some, like torrent beetles and stonefly larvae, have low-slung or flattened bodies that enable them to cling to rocks without being washed away.

Water quality often determines which species will be supported in a particular river or stream. Temperature, velocity, oxygen content, mineral content, and muddiness are all factors. Cold-water trout, for example, prefer cool, shady streams, while snails require calcium-rich waters in order to build their shells. Estuaries are a combination of both salt- and freshwater environments. They are home to species especially adapted to those conditions.

Microorganisms Microorganisms cannot be seen by the human eye without the use of a microscope. Those found in rivers and streams include the transparent stentor. The larvae of animals whose adult forms may be larger, such as dragonflies, worms, and frogs, may spend some time as microorganisms living in the water.

Bacteria Bacteria are microorganisms found throughout rivers and streams and make up much of the dissolved matter in the water column. They provide food for smaller animals and help decompose (break down) the dead bodies of larger organisms. For those reasons, their numbers increase along the banks where the most life is found.

Invertebrates Invertebrates are animals without a backbone. They range from simple flatworms to more complex animals such as spiders and snails.

Many species of insects live in rivers and streams. Some, like the diving beetle, spend their entire lives in the water. Others, like black flies, live in the water while young but leave it when they become adults.

River Blindness

In tropical Africa, South America, and Central America, a round worm causes a disease in humans called river blindness. The worm is introduced by the black fly, which thrives around fast-moving rivers. The fly carries the larvae of the worm and spreads them to humans in its bite. The larvae burrow under the skin and spread throughout the body, eventually reaching the eyes. In some African villages, 15 percent of the people are infected.

Mollusks, such as freshwater mussels, are invertebrates with a hard outer shell that often inhabit rivers.

Common river and stream invertebrates Common invertebrates found in rivers and streams include diving beetles, mayflies, water fleas, torrent beetles, stoneflies, dobsonflies, water boatmen, water striders, whirligig beetles, black flies, fisher spiders, water scorpions, water sticks, leeches, flatworms, snails, freshwater clams, freshwater mussels, raft spiders, caddisflies, and freshwater crayfish.

Food Insects may feed underwater as well as on the surface. They may be herbivores (plant eaters), carnivores (meat eaters), or scavengers that eat decaying matter.

The diets of invertebrates other than insects varies. Some snails are plant feeders that eat algae, while freshwater crabs are often omnivorous, eating both plants and animals.

Reproduction Most invertebrates are insects, which have a four-part life cycle. The first stage is spent as an egg. The second stage is the larva. It may be divided into several stages between which there is a shedding of the outer skin casing. A caterpillar is an example of an invertebrate in the larval stage. The third stage is the pupal stage, during which the insect lives in yet another protective casing, like a cocoon. Finally, the adult breaks through the casing and emerges.

For some insects, their entire life cycle is centered on breeding. Mayflies, for example, live for an entire year in streams as larvae. After they shed the larval casing, they breed, lay eggs, and die; all within a day or two. If enough mayflies survive the larval stage, the numbers of dead adults can be enormous. On one occasion, a snowplow had to be used to remove a layer of mayflies several feet thick from a bridge across the Mississippi.

Raft spiders live along the edges of slow-moving rivers where they are able to run across the water's surface in search of insect prey. Bristles on their feet help distribute their weight so they do not sink. Raft spiders can be identified by two yellow stripes running the length of their bodies.

Adult caddisflies are creatures that come out only at night, rarely feed, and live for only a couple of days. The larvae of caddisflies use stones, sand, shells, and other items to build a protective shell around themselves. The only opening is a hole at the front through which the larva's head emerges to feed. Some larvae eat algae, and others build nets to catch food carried by the current.

The 540 species of freshwater crayfish are found in rivers on all continental landmasses except Africa. The crayfish is related to the lobster and needs mineral-rich waters to build its protective shell. Species range from 1 inch to 16 inches (2.5 to 40 centimeters) in length. They seek shelter under stones in the stream bed or may even burrow into it, sometimes as far as 20 feet (6 meters) down. They catch prey, such as small fish, with their large pincer claws. Some blind, albino (colorless) species live in underground rivers.

Spotted newts hatch from eggs in the water, and then live part of their adult lives on land.
IMAGE COPYRIGHT STEPHEN BONK, 2007. USED UNDER LICENSE FROM SHUTTERSTOCK.COM.

Amphibians Amphibians, including frogs, toads, newts, and salamanders, are vertebrates, which means they have a backbone. Amphibians live at least part of their lives in water. They must usually remain close to a water source because they breathe through their skin, and only moist skin can absorb oxygen. If they are dry for too long, they will die.

Amphibians are cold-blooded animals, which means their body temperatures are about the same temperature as their environment. As temperatures grow cooler, they slow down and seek shelter in order to be comfortable. In cold or temperate regions, some amphibians hibernate (remain inactive), digging themselves into the mud. When the weather gets too hot, many go through another similar period of inactivity called estivation.

Common river and stream amphibians The most common amphibians found in rivers and streams all over the world are salamanders, frogs, toads, and newts.

The palmate newt lives on land for part of the year where it hunts at night for worms and other small animals. In spring it returns to the water where it breeds and lays eggs. Most newts lay eggs one at a time and attach them to water plants. Some species even wrap each egg in a leaf for protection.

Walking on Water

The Jesus Christ lizard, or basilisk, of Central America got its name from its ability to walk on water. Scales and a flap of skin on its hind toes increase the surface area of its feet and enable it to scamper on the water's surface. Being lightweight also helps.

Another water-walker, or one that appears to be, is the jacarana, or lily trotter, bird of the tropics. Jacaranas have extremely long toes that distribute their weight over the surface, but they are actually walking on the floating leaves of water plants.

Food   In their larval form, amphibians are usually herbivorous, while the adults are usually carnivorous, feeding on insects, slugs, and worms. Those that live part of their lives on land have long, sticky tongues with which they capture their food.

Reproduction   Most amphibians lay jelly-like eggs in the water. Frogs, depending on the species, can lay up to as many as 50,000 eggs, which float beneath the water's surface. After being fertilized by the sperm of male frogs, the eggs hatch into tadpoles (larvae) and require a water habitat as they swim and breathe through gills. When the tadpole turns into a frog, it develops lungs and can live on land.

Spotted newts hatch in the water and live there as larvae, even developing gills. Later, they lose their gills and live for a time on dry land. Two or three years later, they return to the water where they live the remainder of their lives.

Some amphibians, like the common frog, prefer quiet water for breeding. Others, like the large hellbender salamander, seek rushing streams.

Reptiles   Reptiles are cold-blooded vertebrates that depend upon the environment for warmth. They are more active when the weather and water temperature become warmer. Many species of reptiles, including snakes, lizards, turtles, alligators, and crocodiles, live in temperate and tropical rivers and streams. All are air-breathing animals, with lungs instead of gills, which have adapted to life in the water. The Southeast Asian fishing snake, for example, can close its nostrils while swimming submerged.

Many reptiles go through a period of hibernation in cold weather because they are so sensitive to the environment. Turtles, for example, bury themselves in the mud. They barely breathe and their energy comes from stored body fat. At the other extreme, when the weather becomes very hot and dry, some reptiles go through estivation.

Common river and stream reptiles   Common reptiles found in rivers and streams include the water moccasin, the viperine water snake, the eastern water dragon, the soft-shelled turtle, the matamata turtle, the yellow-bellied terrapin, the snapping turtle, the anaconda, and the crocodile.

The anaconda of South American rivers is the world's largest snake, some individuals attain lengths of more than 33 feet (10 meters). Although they are excellent swimmers and prefer to hunt along the edges of rivers and swamps, they can also climb trees. Anacondas are constrictors, killing their prey by coiling their body around it and squeezing it. Their young are born alive in the water.

Crocodiles are fierce predators that spend time both in and out of the water. IMAGE COPYRIGHT JOSHUA HAVIV, 2007. USED UNDER LICENSE FROM SHUTTERSTOCK.COM.

Crocodiles hunt at twilight in tropical rivers. During the day they bask in the sun at the river's edge. As the light dims, they seek their prey. Some, such as the New Guinea crocodile, eat just fish. Most other species eat a variety of foods. They swallow smaller prey whole, but larger animals are dragged underwater where they drown and are torn apart and eaten.

Food All snakes are carnivores. Water snakes eat frogs, small fish, and crayfish. Some species of turtles are omnivores.

Reproduction The eggs of lizards and turtles have either hard or rubbery shells and do not dry out easily. Most are buried in warm ground, which helps them hatch. The eggs of a few species of lizards and snakes are held inside the female's body, and the babies are born alive.

Crocodiles keep their eggs warm in nests that can be simple holes in the ground or constructions above the ground made from leaves and branches.

Fish Like amphibians and reptiles, fish are cold-blooded vertebrates. There are two types of freshwater fish. The first type, the parasites, which include some species of lampreys, attach themselves with suckers to other animals and suck their blood for food. The second type, which eat plant foods or catch prey, are the most numerous. They use fins for swimming and breathe with gills.

Many species of fish, such as archer fish, swim in schools. A school is a group of fish that swim together in a coordinated manner to discourage predators.

Food Some fish, such as the grass carp, eat plants while others, such as bluegills, depend upon insects, worms, and shellfish. Larger fish, such as the sturgeon, often eat smaller fish, and a few species feed

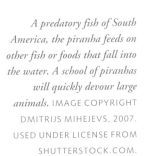

A predatory fish of South America, the piranha feeds on other fish or foods that fall into the water. A school of piranhas will quickly devour large animals. IMAGE COPYRIGHT DMITRIJS MIHEJEVS, 2007. USED UNDER LICENSE FROM SHUTTERSTOCK.COM.

on carrion (dead animals). Most fish specialize in what they eat and where they find their food. Some feed on the surface, and others seek food in deep water. Some prefer rushing streams and others calm pools. The arawana of Southeast Asian rivers has a large, upward-pointing mouth that helps it feed on the surface.

Common river and stream fish   Typical fish found in rivers and streams include the roach, rudd, tench, bream, perch, gudgeon, carp, bitterling, African knifefish, elephant fish, hatchetfish, shovel-nosed sturgeon, white sturgeon, paddlefish, freshwater shark, arapaima, stickleback, piranha, catfish, salmon, and trout.

Piranhas (purr-AÑ-ahs) are predatory fish of South America, known for their several rows of sharp, triangular teeth. They usually feed on other fish, or fruits and seeds that fall into the water. They become particularly dangerous during the dry season when the water level drops and fish gather together in schools. A school of piranhas can attack and devour large animals, including humans.

Many species of catfish inhabit rivers worldwide. Some tropical varieties can climb out of the water and walk overland from one river or stream to another. They range in size from the small species that live in mountain streams, to those that live in the Danube River in Europe and grow to a length of 13 feet (4 meters) and a weight of 400 pounds (180 kilograms).

Catfish got their name from the long feelers around the mouth that resemble cats' whiskers. They have poisonous spines along the fins that can cause painful injuries. Catfish adapted to life in fast-moving streams have fins with suckers or large, fleshy lips that help them cling to rocks and even climb. Bullheads, a species of catfish, have flattened bodies, which enables them to squeeze under stones and resist the current. Some freshwater catfish communicate with each other through grunting and clicking sounds.

Salmon live in both fresh and salt water. The young hatch in fast-moving rivers and streams, where they live for about the first three years of their lives. They then migrate to the sea where they live for up to four years, feeding on small fish and shellfish. When it comes time to breed, they return to the river of their birth, swimming upstream. Salmon are called *anadromous* fish, which is Greek for "running upward." After breeding, they die; their decaying bodies provide an injection of nutrients to their birth area.

In general, trout are found in oxygen-rich, cool, fast-moving streams with beds of gravel. They are carnivorous, feeding on insects, freshwater shrimp, and clams. Some species also prey on other fish. Rainbow trout, which originated in North America, have a silver belly and an iridescent stripe along the side. The Aurora trout, a subspecies of the brook trout, is a spectacularly colored fish. They are native to two small lakes in Ontario, Canada. In the 1960s, the Aurora trout was almost driven to extinction due to acid rain. The lakes were treated with lime to decrease the acidity of their waters and the trout was reintroduced.

Reproduction   Most fish lay eggs, and many species abandon the eggs once they are laid. Others build nests and care for the new offspring until they hatch. The male stickleback lures several females to his nest where they lay their eggs. He then fertilizes the eggs and guards them until they hatch. Still other species, such as certain catfish, carry the eggs with them until they hatch, usually in a special body cavity or even in their mouths.

Shocking River Residents

There are lots of good reasons to not swim in tropical rivers, and one is to avoid an electric shock. The African catfish and the Amazonian eel are two tropical river residents that produce their own electricity and use it as a weapon.

The body of an electric eel functions like a chemical battery. Thousands of electricity-producing cells are stacked in columns and oriented in the same direction. When the eel sights something tasty, a nerve impulse from the eel's brain triggers the flow of electric current from its tail, through the water, toward its head. When the eel touches its prey, the circuit is interrupted and the current flows into the body of the victim, stunning it. An adult eel can produce 600 volts of electricity, which is enough to knock out an animal the size of a horse or a human.

The Fish that can Climb Ladders

In order to breed, Atlantic salmon return from the sea to rivers along the Atlantic and Pacific coasts where they first hatched. Most of these rivers have been dammed in places, preventing the fish from continuing on their journey upstream. To aid them, some states have built "fish ladders," water-covered steps alongside the dams that allow the fish to jump from one level to the next as they would over a natural waterfall. In spite of this technological aid, the numbers of salmon continue to decline, as do the numbers of animals who depend upon the salmon for food.

Birds   Birds are vertebrates, and many different species live near rivers and streams. These include many varieties of wading birds, waterfowl, and shore birds. Most visit in search of food, and certain species use riverbanks as nesting places.

Most aquatic birds prefer quiet waters, but a few are specially adapted for the speed and turbulence of life upstream. One example is the torrent duck of the Andes Mountains in South America. Built like an ordinary duck in most respects, it has a longer tail to help it steer when swimming in fast currents, and claws on its webbed feet that allow it to cling to rocks.

Common river and stream birds   Birds found around rivers and streams include mergansers, warbles, reed buntings, ospreys, sunbitterns, finfoots, egrets, and wood ducks. Kingfishers, herons, fish eagles, ibises, grebes, and cranes are the few bird species that spend their entire lives around freshwater habitats. They fall into four basic categories: wading birds, shorebirds, waterfowl, and birds of prey.

Wading birds, such as herons, have long legs and wide feet for wading through shallow water. They have long necks and bills necessary for nabbing food. There are approximately 100 species of heron.

Shorebirds, such as the sandpiper, feed and nest along the banks and prefer shallow water. Some, like the godwit, have long, slender, upturned bills that help them sift through mud in search of food. Other birds, like the ruddy turnstone, have bills that are curved to one side, which helps them to overturn pebbles. The red knot is an example of a shorebird that has specialized bill cells that are pressure sensitive, allowing them to detect shellfish buried beneath the sand.

Waterfowl are birds that spend most of their time on water, such as ducks, geese, and swans. Their flattened bills are designed for grabbing vegetation, such as sedges and grasses, on which they feed. Waterfowl are strong fliers and swimmers. Of the 150 species, most prefer freshwater environments. Migrating waterfowl play a major role in dispersing small aquatic organisms that attach to their feet or feathers.

Osprey and fish eagles are birds of prey that only hunt in freshwater. Other birds of prey include gulls, terns, and stilts.

Wading birds have long legs and wide feet for walking through shallow water, and long necks and bills for grabbing food. IMAGE COPYRIGHT EVGENY V. KAN, 2007. USED UNDER LICENSE FROM SHUTTERSTOCK.COM.

Food   Nearly all birds must visit a source of fresh water each day to drink, and many feed on aquatic vegetation or the animals that live in the water. The dipper can actually walk underwater in search of small animal prey.

One of the adaptations of birds to aquatic life is their beak, or bill. Some are shaped like daggers for stabbing prey such as frogs and fish; others are designed to root through mud in search of food. Water-dwelling birds spend much time preening their feathers because they depend on their feathers to keep them dry.

Reproduction   All birds reproduce by laying eggs. Male birds are often brightly colored to attract the attention of females. After mating, female birds lay their eggs in a variety of places and in nests made out of many different materials. One parent usually sits on the nest to keep the eggs warm while the other searches for food.

Mammals   Mammals are warm-blooded vertebrates covered with at least some hair, that bear live young and nurse their young with milk. Aquatic mammals, such as the muskrat, often have fur that is waterproof. They may have webbed toes for better swimming. The toes of the duck-billed platypus of Australia, for example, are webbed.

Food   Certain aquatic mammals are carnivores. The water shrew, for example, eats insects, worms, and frogs. Others, like muskrats, are omnivores, while beavers are herbivores.

What Am I?

The duck-billed platypus, which is found in Australian rivers, appears to be a cross between a reptile, a duck, a beaver, and an otter. It has a ducklike beak, webbed feet, and lays eggs. It also has dense fur and a flat, beaverlike tail, and suckles its young with milk. The platypus is a good swimmer and dines on the river's insect population. Officially, the platypus is a mammal and cousin to the anteater. Some scientists believe that the species originated during the evolution of reptiles into mammals.

Reproduction Mammals give birth to live young that have developed inside the mother's body. Some mammals, such as otters, are helpless at birth, while others, like dolphins, are able to walk or swim immediately. Many are born with fur and with their eyes and ears open. Others, like baby muskrats, are born hairless and blind. Mammals feed their young with milk produced by the mother.

Common river and stream mammals Mammals found in and around rivers and streams include manatees (sea cows), capybaras (large rodents), tapirs, minks, water shrews, beavers, dolphins, otters, and hippopotami.

Of the five recognized species of freshwater dolphins, four live in rivers. River dolphins are found in the Ganges, Indes, Amazon, and Yangtze rivers. They are almost blind and find their way through muddy waters by sending out sound waves that bounce off obstacles to warn them of what's ahead and help them find the fish on which they feed. Like their saltwater cousins, freshwater dolphins are considered intelligent and some fishermen use them to herd schools of fish into waters where they are more easily caught. They differ from their marine cousins in that they are slower swimmers and have larger heads.

Dolphins produce a single calf that is capable of swimming and breathing within the first minutes of its life. Some mothers have been observed guiding the calf to the surface, as if to help it reach air. River dolphins are threatened by hunting, entanglement in fishing nets, and pollution.

Of the thirteen species of otters, twelve are found in freshwater. River otters have webbed feet and strong tails that make them good swimmers. Their thick, water repellent fur coats keep them warm and dry. They feed on frogs, birds, fish, and small mammals. Their eyes, ears, and nostrils are placed high on their heads, which enables them to look sharply around while swimming. They dig dens into the riverbank where they care for their young until the young are able to live on their own. Otters are playful animals and spend much of their time at acrobatic games. Weasels, skunks, and badgers are relatives of the otter that are found in rivers and streams.

River otters spend most of their time on land, but are excellent swimmers. They eat fish and other aquatic organisms.
IMAGE COPYRIGHT JOHN WILLIAMS, 2007. USED UNDER LICENSE FROM SHUTTERSTOCK.COM.

The name hippopotamus means "river horse" in Greek, and, although hippos look ungainly, they are fast both in the water and on land. They graze on vegetation on land at night and spend their days dozing in the muddy waters of African rivers. Adult males may reach a length of 16 feet (5 meters) and weigh as much as 4 tons (3.6 metric tons). Their skin is nearly hairless, and the eyes and nostrils protrude so that they are above the water even when the rest of their body is submerged. There are two species of hippo. The larger, common species is widespread in Central Africa. The scarcer, pigmy hippo is restricted to Western Africa.

Endangered species   Seven types of salmon and two species of trout are on the endangered list in the northwestern United States. The Topeka shiner and the Arkansas River shiner are threatened in the central United States and the snail darter in Tennessee. As rivers and streams become polluted, many species of birds, such as the fish eagle, have become threatened. Birds cannot use the polluted waters for food or as nesting areas, which causes their numbers to decline.

At one time hippos were found in almost all African rivers. As the human population has grown, their habitat has shrunk and they are now found in only a few rivers. The Amazonian manatee, the only species of freshwater manatee, is now protected against hunters, who kill it for its meat.

River of Death

In Greek mythology, the River Styx traveled through the underworld, the land of the dead. The water of that river was supposedly so poisonous that it would dissolve any vessel used to carry it except one made from a horse's hoof. An ancient account claims that Alexander the Great (356–323 BC), king of Macedon, was poisoned by Styx water. There may be a bit of truth in the claim; some experts believe he died of a bacterial infection caught by drinking water from the Euphrates River, which carried raw sewage out of the ancient city of Babylon in what is now southern Iraq.

Human Life

Rivers have always been intimately connected to human life, and river valleys have been the birthplaces of the world's great civilizations. The Nile River is the scene of the oldest and perhaps the most remarkable of ancient civilizations, Egypt. The Nile's yearly floods brought rich sediments to the soil, and the people learned to divert the floodwaters for agriculture and drain the swamps. Chinese civilization got its start in the lower valley of the Huang He, where many floods and extreme weather forced people to develop the technology necessary for life to continue successfully there. In the valley of the Tigris and Euphrates rivers, Sumerian civilization developed as people fought to clear jungle swamps and manage floodwaters. The same was true of the Indus civilization, which arose in the valleys of the Ganges and Indus rivers in India and Pakistan. While rivers offered opportunities for water and transportation, they also presented challenges that inspired people to excel in order to survive.

Impact of rivers and streams on human life   People use rivers and streams for water and food; boundaries, transportation, and trade; recreation and building sites; and creating hydroelectric power.

Water and food   Many rivers and streams are sources of drinking water and for such things as bathing and laundry. In developed countries, about half the freshwater supply is used by industry. In less wealthy countries, up to 90 percent is used for irrigation (watering) of crops. The earliest irrigation systems were built on the banks of the Nile River about 6,500 years ago. Pumping systems and waterwheels soon followed. About 642,473,992 acres (260,000,000 hectares) of land is now irrigated worldwide.

The fish and plants in rivers and streams are often a source of food for humans. Although most fish used for food come from the ocean, commercially important freshwater fish include trout and salmon. Farmers often use aquatic plants, such as marsh grass, reeds, and sedges, for feeding livestock.

Boundaries, transportation, and trade
Throughout history, rivers have acted as boundaries. If they were too big or too fast to cross easily, they limited overland travel. During exploration of the American West, for example, wide rivers like the Missouri meant that travelers had to build a boat or raft or find a spot where the channel was narrow and the water shallower, so that horses could swim across.

Rivers can act as boundaries between two different territories. Many countries use rivers to help define their borders. The Rio Grande separates Texas from Mexico, and the Congo in Africa divides the Democratic Republic of the Congo and Zaire.

Rivers have always provided a means of transportation and trade. Cities built at the edge of rivers could import or export goods upstream or downstream or even to sea more easily and less expensively than cities reachable only by land routes. An example is the Mackenzie River of Canada's Northwest Territories, which flows into the Arctic Ocean. Early European settlements, such as Fort Good Hope, were built along its shores as part of the fur trade. It is now used to transport petroleum and uranium ore.

Swan Song for the Thames

Even before 1800, the Thames (TEMS) River in England was polluted by chemicals and other materials used in manufacturing. With the invention of flush toilets, things got much worse. Where did the sewers empty out? In the Thames. The fish suffocated to death and swans and other waterbirds left for a healthier part of town. Until the 1950s, the Thames was considered one of the most polluted rivers on the planet.

Public protest finally grew so loud that things began to change. Strict regulations were created to limit what could be dumped into the water. Sewage-treatment plants were improved, and special equipment was installed in some locations to mix the water with oxygen. Before long, the river began to recover. Within thirty years, pollution had been reduced by 90 percent, and oxygen content rose to healthy levels. The fish returned, and so did aquatic plants. Even the swans moved back.

Many rivers are still important routes for bulk cargo, such as coal. The Chang Jiang River in China, for example, is navigable by ocean-going freighters for 680 miles (1,090 kilometers) inland and the St. Lawrence Seaway between the United States and Canada for about 2,300 miles (3,800 kilometers).

Recreation and building sites   Streams and even some rivers are popular for sport fishing and boating, but pollution makes most of them undesirable for swimming. In urban regions, the areas along riverbanks have become popular sites on which to build homes. Some rivers and streams may be diverted to create artificial lakes or ponds in order to add beauty and wildlife to a residential area.

Hydroelectric power and other resources   When river waters rush over the blades of a turbine (energy-producing engine), the blades turn

The African Queen

A mixture of comedy, adventure, and romance, *The African Queen* is a classic film about two people during World War I (1914–18) who take an ancient, ramshackle boat up a treacherous river in Africa in hopes of finding and destroying an enemy ship. They encounter rapids, leeches, hippos, and other river dangers during the wild ride to their destination. The film starred Katharine Hepburn and Humphrey Bogart and won the Academy Award in 1951. It was based on the book of the same name by British author C. S. Forester (1899–1966).

quickly. Problems occur when the volume of sewage is so large and bacteria use up so much oxygen that plants and animals cannot survive. Additionally, this process results in the formation of ammonia, which is poisonous to most animals. Sewage fungus, a combination of bacteria and molds, may grow on the surface of the water.

Another source of river pollution is industrial chemicals and heavy metals, which may dissolve in the water or remain as solids. Enormous quantities of these materials poison animals, clog gills in fish, and bury the riverbed in sediments. Industries are responsible for dumping heated water (thermal pollution) into rivers, raising their temperature and killing many species of plants and animals that require a cooler environment.

Shipping can cause pollution. Ships discharge waste into the water or develop leaks and spills that threaten the water's purity and destroy wildlife.

Eutrophication (yoo-troh-fih-KAY-shun) occurs when fertilizers used in farming get into rivers and streams, spurring a greatly increased growth of algae in slow-moving waters. These plants form a thick mat on the surface and block the sunlight, causing submerged plants to die. As the dead plants decay, oxygen in the water is used up, killing fish and other plant life.

Flood control   Flooding nourishes the soil in floodplains, creates natural wetlands, and supports wildlife. Floods can also be destructive and dangerous to human life. For this reason, efforts have long been made to control floods using dams and artificial levees. Many of these changes have not only harmed wildlife, they have sometimes caused more problems than they have helped. The Aswan High Dam in Egypt is an example. Completed in 1970, the dam is 2.5 miles (4 kilometers) in length and rises 364 feet (111 meters) above the Nile riverbed. The dam was built in order to irrigate desert regions. However, dissolved salts have built up in the irrigated areas and increased the saltiness of the Nile itself. The Nile used to carried vast amounts of sediments to the Mediterranean. With their reduction, the sardine fishery in the eastern Mediterranean has been destroyed. The dam has increased erosion in the Nile waterway and in the delta region.

Giants of the River

The last part of the twentieth century has seen the building of colossal dams designed to bring hydroelectric power and flood control to undeveloped countries.

As of 1999, the world's largest was the Itaipu Dam on the Parana River between Brazil and Paraguay in South America. Completed in 1991 at a cost of over $20 billion, it has eighteen huge turbines that generate as much as 12,600 megawatts of electricity. (By comparison, the largest dam in the United States, the Grand Coulee, outputs only 9,700 megawatts.) Over 30,000 workers took sixteen years to create the 4.8-mile- (7.6-kilometer-) long, 500-foot- (152-meter-) high structure, using enough concrete to build eight medium-sized cities. Its reservoir covers 870 square miles (2,262 square kilometers), and thousands of farm families had to be relocated. The presence of the reservoir has caused some cooling of the local climate, which endangers wheat and other commercial crops, but the dam produces clean, renewable energy and creates thousands of jobs.

Soon to overtake the Itaipu in scope and power production is the Three Gorges Dam on the Chang Jiang (Yangtze) River of China. With its twenty-six turbines, the Three Gorges is expected to output 18,200 megawatts of electricity—as much as eighteen nuclear plants—and help bring prosperity to central China. At a projected cost of $75 billion, the dam is scheduled for completion in 2010 and will be 610 feet (187 meters) high and 6,864 feet (2,092 meters) wide. Its reservoir will be 370 miles long (592 kilometers), filling the towering limestone canyons along the river's route. As many as fourteen towns will be drowned and nearly two million people relocated. There is opposition to the dam because it will kill many aquatic animals and destroy ecosystems and valuable archaeological sites. On the other hand, it is hoped that the dam will end the disastrous flooding common to the Chang Jiang, which has killed as many as 300,000 people during the past century alone.

Other giant dams in progress include the Ataturk Dam on the Euphrates River in Turkey and the Sardar Sarovar in India, part of a proposed thirty-dam project on the Narmada River.

In Quebec, Canada, many people are opposing the James Bay Project, a proposed hydroelectric dam that will alter the region. People are beginning to wonder if it may be wiser to allow rivers to flow unobstructed.

Wild and Scenic Rivers Act   In 1968, the U.S. Congress passed the Wild and Scenic Rivers Act, which established a program to study and protect free-flowing rivers. The act prohibits building hydroelectric or other water development projects on certain rivers. In order to qualify for protection, a river must be undammed and have at least one outstanding resource, such as a wildlife habitat or historic feature. As of 2007, 130

This vast floodplain is underwater for several months each year, and many channels and oxbow lakes form. In Obidos, Brazil, its depth is about 300 feet (91 meters). The maximum width of its permanent bed is 10 miles (16 kilometers). The width of its mouth is 360 miles (580 kilometers). No delta has formed because the 3,000,000 tons (2,721,554 metric tons) of sediment discharged daily are carried northward by ocean currents and deposited along the coast of French Guiana.

The Amazon basin supports the largest area of rain forest in the world and the greatest plant and animal diversity, with about one million species. Plants, such as cassava, tonka beans, guava, and calabash, are found here. Trees include palms, myrtles, laurels, acacias, and rosewoods.

The Amazon region hosts more than two million species of insects. The number of Amazon fish species has been estimated at 1,500, more than 18 percent of the world's known species. A few, such as the piaraucu, are commercially important. Most streams are home to piranhas, sting rays, and giant catfish.

Turtles and giant constrictor snakes represent reptiles. Birds live among the trees, and some, such as the fish eagle, hunt food in the river. Mammals include the capybara. The giant otter, which can grow to more than 6 feet (1.8 meters), is heavily hunted and almost extinct. It is only found in the Amazon.

Most human inhabitants of the Amazon basin have been Indians, such as the Jivaro and the Yanomamo, but the area has never been heavily populated. The soil is poor and not really suitable for farming, although huge tracts of forest are being destroyed for agricultural purposes. The river itself has been used for transportation and trading since the earliest times, and ocean-going vessels can travel as far as 1,000 miles (1,600 kilometers) upstream to Manaus, a city in northern Brazil.

Nile River

Location: Egypt, Burundi, Rwanda, Uganda, Ethiopia, Kenya, Tanzania, and Sudan in Africa

Length: 4,160 miles (6,695 kilometers)

Discharge Basin: 1,100,000 square miles (2,860,000 square kilometers)

The Nile The Nile is the longest river in the world, at a length of 4,160 miles (6,695 kilometers). Its main tributaries are the White Nile, which has its headwaters in Burundi, and the Blue Nile, which rises in the highlands of western Ethiopia. Lake Albert, Lake Victoria, and Lake Tana all contribute water. The Blue Nile and the White Nile join together at Khartoum in Sudan and continue north into southern Egypt. Six stretches of rapids and waterfalls break its flow, ending at Aswan. Between Aswan and the Mediterranean Sea, the river creates a wide, rich floodplain that has been cultivated for more than 3,000 years. The Nile delta lies north of Cairo, a 100-mile- (160-kilometer-) long stretch of rich silt deposits cut by many streams. This area provides almost all of Egypt's food crops, and cotton, one of the country's major exports.

Precipitation varies along the Nile's length, but most of its basin receives no rain from November to March. At Khartoum, 5 inches (12.7 centimeters) of rain falls annually, but Cairo receives only 1 inch (2.5 centimeters). The far south receives 79 inches (200 centimeters) of rain each year. Before the Aswan High Dam was completed in 1970, heavy rains in the south caused floods. The floods were vital to agriculture in the region, but they often caused serious damage. The dam prevents flooding, but holds back the silt, so the land has to be fertilized by other means. The dam has also led to erosion of the delta.

Grasslands and tropical forests occur where the river passes through Ethiopia and eastern Africa. Papyrus, water lettuce, sedges, and water hyacinth grow in and around the river. Farther north, drier conditions prevail and support only acacia and scrub. Beyond that, the Nile passes through true desert. In Egypt, desert land along the Nile is irrigated and cultivated.

The river supports fish, snakes, turtles, crocodiles, and lizards. In its southern waters, hippopotami can be found.

About 3100 BC, the ancient Egyptian civilization began on the banks of the Nile and was totally dependent upon the river both for water and for the sediments that enriched the soil. Lower Egypt, which was centered around the delta, was fertile and green. Upper Egypt was hot and dry, and the band of land suited for farming was only a few miles wide. Crops of all kinds were grown, and cattle, goats, pigs, and sheep were pastured. The river provided fish and waterfowl, and papyrus and other reeds and grasses were turned into fibers for baskets, boats, and paper. River mud was used to make bricks and, over time, the Egyptians developed great architectural skill, creating the magnificent pyramids that still stand.

Almost 97 percent of Egypt's population still lives in the Nile valley and delta. Fishers and farmers live in the irrigated north and in southern regions where rainfall is adequate. In northern Sudan, nomads raise cattle and camels. More arid (dry) regions are extremely dependent upon the river for their economy. The Aswan High Dam produces electricity and has increased irrigated land by 30 percent.

The Huang He (Yellow River) The Huang He (wang HO) or Hwang Ho, which is Chinese for Yellow River, is the second longest river in China and flows eastward from the Kunlun Mountains through the

Huang He (Yellow) River
Location: China
Length: 2,901 miles (4,641 kilometers)
Discharge Basin: 486,000 square miles (1,263,600 square kilometers)

central plains to the Yellow Sea. Its waters are clear in the upper regions, but as it passes through Shanxi and Shanxi provinces, it picks up large quantities of sediments—1,600,000,000 tons (1,440,000 metric tons) annually—that give it its yellow color and, therefore, its name. These sediments made the Huang He's delta the fastest growing in the world. By the 1990s, dam construction, particularly Sanmen Gorge in the Henan Province, is causing the delta to erode.

Average precipitation in the Huang He basin is about 16 inches (40 centimeters) a year, and average annual discharge is 11 cubic miles (49 cubic kilometers). Rainfall varies greatly. In winter and early summer, the river is at its low point. In spring and late summer, water levels are high, sometimes as much as 70 feet (21 meters) above its banks, and severe flooding is common. The Huang He has overflowed its dikes (retaining walls) hundreds of times in the past 3,000 years, causing so much damage and so many deaths that it has been referred to as "China's Sorrow." In an attempt to control flooding, artificial embankments have been created along 1,120 miles (1,800 kilometers) of the river's length.

Much of the Huang He basin is grassland. Cattail marshes are home to many geese, ducks, and swans. Carp and catfish live in the river itself.

China's great civilization, dating from at least as early as 2000 BC, probably began in the Huang He basin. The basin supports a population of more than 120 million people. In many places, the current of the Huang He is so swift and the channel so shallow that it is of little use for navigation, but the floodplain is developed for agriculture.

The Chang Jiang (Yangtze)   The Chang Jiang (zhang jee-ANG), also spelled Yangtze, is the longest river in China at 3,915 miles (6,300 kilometers). It crosses the central region from east to west, where it flows into the East China Sea near Shanghai. Its headwaters are in the Tibetan highlands at elevations of about 18,000 feet (5,480 meters), and its course makes many bends. Below Wan-hsien it travels about 200 miles (320 kilometers) through spectacular gorges and deep canyons where rapids form. In the region of the gorges, the river is 500 feet (152 meters) deep, making it the world's deepest. During summer floods, the water in this region can rise 200 feet (61 meters) and travel at velocities faster than 13 miles (21 kilometers) per hour. Boats are often towed through the gorges by large gangs of men on shore using 0.5-mile-long (0.8-kilometer-long) ropes made from bamboo.

Chang Jiang (Yangtze)
River
Location: China
Length: 3,915 miles (6,300 kilometers)
Discharge Basin: 756,498 square miles (1,966,895 square kilometers)

Until 1957, there was only one bridge across the Chang Jiang, near Wuhan. Here, the river forms a vast floodplain, and flooding often occurs in summer. Since that time sixteen other bridges have been built. In 1998, the worst flood in forty-four years caused 3,004 deaths and twenty billion dollars (U.S.) in damages. More than fourteen million people were left homeless.

Precipitation averages 24 inches (60 centimeters) annually. Average discharge into the sea is 1,010,000 cubic feet (28,600 cubic meters) per second, and 250,000,000 tons (336,796,185 metric tons) of sediment is deposited annually.

Deciduous forests cover those areas of the basin that are not cultivated for farming. About 70 percent of China's rice crop is grown in the Chang Jiang basin, and about 300 million people live in the region.

The Chang Jiang River has been home to the giant paddlefish, which may now be extinct. The rare Chang Jiang river dolphin was declared extinct in 2006, due to pollution, dam construction, hunting, and depletion of its food supply. Animal life varies along its length and includes thrushes, pheasants, and antelopes.

The Chang Jiang is important commercially. The volume of transportation is greater than all the other waterways of China combined. About 25,000 miles (40,000 kilometers) of main river and tributaries can be navigated by small craft, and ocean-going ships can reach Wuhan, a city in central China. Five of China's largest cities—Shanghai, Wuhan, Chongqing, Chengdu, and Nanjin—are on or near the Chang Jiang system.

The river is considered ideal for generating hydroelectric power, and the gigantic Three Gorges Dam, which began construction in 1994, is predicted to provide at least 15 percent of China's electricity needs. The dam will also help control flooding, but many scenic areas will be permanently covered by its reservoir. Thousands of people are being forced to relocate, and some scientists believe the dam will have undesirable effects on the environment after it is completed around 2010. Once completed, the dam will be the world's largest at 6,864 feet (2,092 meters) wide and 610 feet (186 meters) high.

The Ganges (Ganga)   The Ganges (GANN-jeez) crosses northern India and then flows south through the provinces of Uttar Pradesh, Bihar, West Bengal, and finally Bangladesh. Its headwaters are the Alaknanda and Bagirathi Rivers, which begin in the Himalaya Mountains. Its delta

Ganges (Ganga) River
Location: India and Bangladesh
Length: 1,557 miles (2,491 kilometers)
Discharge Basin: 188,800 square miles (490,880 square kilometers)

begins 200 miles (320 kilometers) from the sea, where it joins with the Jumna River to form the Padma and then flows into the Bay of Bengal, where it discharges more sediment into the sea than any other river system.

The flow of water in the Ganges varies with the seasons, but the changes are seldom violent. It travels through dry plains, drawing water from several tributaries such as the Jumna, Gogra, Gandak, Kasi, and Brahmaputra. In the last 1,000 miles (1,600 kilometers), flow is sluggish. Floods are common, especially in the delta region where tropical storms and tidal waves (giant waves associated with earthquakes) occur. During one such storm in 1970, about one million people were killed.

Forests grow on the Gangetic plain. Where rainfall is heavy, evergreens predominate. Carp are found in the river, as are geese and ducks. Mammals include tigers, deer, and wild dogs, but reptiles are more numerous, especially crocodiles, lizards, and snakes. The endangered Bengal tiger makes its home in the Ganges Delta, as does the Indian python and the Asian elephant.

Changing water levels limit the use of the Ganges for large vessels. In the delta region, many places are accessible only by small watercraft. Soils here are rich and fertile, but in the north the land is too high for irrigation unless power equipment is used. Crops include rice, wheat, sugar, cotton, and oil seed, which support 500 million people.

To those of the Hindu faith, the Ganges is holy, and many places along its shore, such as Varanasi (Benares), Allahabad, and Hardwar, have religious significance. Many people go on pilgrimages (religious journeys) to visit these places. Hindus believe that bathing in the Ganges is religiously purifying, and thousands flock to its banks each day. The river is polluted by human sewage, fertilizer runoff, and industrial wastes. In 1985, efforts were begun to clean up the river and are ongoing.

One of the world's earliest civilizations, the Indus, began in the neighboring Indus River valley and spread into the Ganges valley. One of the most well-known archaeological sites is Mohenjo-Daro, beside the Indus River.

Indus River

Location: Tibet, India, and Pakistan

Length: 1,800 miles (2,880 kilometers)

Discharge Basin: 372,000 square miles (967,200 square kilometers)

The Indus The Indus (INN-duhs) River begins in Tibet, travels through part of northern India, and completes its course in Pakistan. Its headwaters are formed by meltwater in the Himalaya Mountains and the Karakoram Range, where its flow is turbulent, unnavigable, and prone to flooding, especially in the summer months. When it leaves the mountains and enters the dry Punjab plains of Pakistan, it broadens and picks up sediments.

The major tributaries of the Indus are the five rivers of the Punjab: the Sutlej, Chenab, Jhelum, Ravi, and Beas. Its delta begins at Thatta and enters the Arabian Sea 70 miles (110 kilometers) farther south, where the river splits into several channels.

Rainfall in the Indus Valley ranges from 5 to 20 inches (13 to 51 centimeters) annually. The Indus is used for irrigation in the drier regions.

Desert vegetation predominates and includes thorn forests and many shrubs. Grasses do not thrive here. Indus baril, Indus gurua, Rita catfish, and snakehead fish are common fish. Sarus cranes and bearded vultures frequent the area, as do crocodiles and cobras. Mammals include deer and wolves.

The Punjab plain is Pakistan's richest farming region, where wheat, corn, rice, millet, dates, and other fruits are grown. Large dams have been built on the river to provide hydroelectric power and water for irrigation. Use of the Indus has been disputed by India and Pakistan as they have tried to determine their borders.

The cities of the great Indus civilization, Harappa and Mohenjo-Daro, were built in the Indus valley around 2500 BC. The civilization appears to have declined gradually. It is believed that barbarian nomads of the Eurasian high plains invaded the region and destroyed what was left of this civilization.

The Colorado   The Colorado River is created by rain and melting snow in the mountains of central Colorado, Wyoming, and Utah. Its main tributaries are the Gunnison, the Dolores, the Green, the San Juan, the Little Colorado, the Gila, and the Virgin. As it travels south and southwest through Utah, Arizona, and Nevada, it creates the magnificent Grand Canyon, cutting deeper into the rock each year. The canyon is 218 miles (349 kilometers) long and, in one spot, 18 miles (29 kilometers) wide and 5,249 feet (1,600 meters) deep. At its mouth, the Colorado empties into the Gulf of California on the Pacific coast.

Discharge rates vary from 3,000 cubic feet (84 cubic meters) per second when water levels are low to 125,000 cubic feet (3,540 cubic meters) per second at flood times.

Colorado River
Location: Colorado, Arizona, Utah, Nevada, and California
Length: 1,440 miles (2,304 kilometers)
Discharge Basin: 244,000 square miles (634,400 square kilometers)

Much of the region is treeless and desertlike, and plants include those that tolerate semiarid conditions. Desert birds and waterfowl are common. Mammals include deer and bears.

The Colorado is one of the most valuable rivers in the world in terms of irrigation and power. The river is prone to flooding, but Hoover Dam and its reservoir (storage area), Lake Mead, have helped prevent disasters. Hoover Dam is the first multipurpose dam, providing both flood control and power generation. The Los Angeles-San Diego region receives much of its water from the Colorado, and arguments have ensued among neighboring states as to how the water should be divided. Environmental concerns have slowed further development of river resources.

Until the Spanish arrived in 1540 and gave the Colorado its present name, which means "red color," the Colorado basin was the home of several Native American tribes, including the Pueblo and the Navajo.

The Congo (Zaire)   The Congo, or Zaire (zy-EER), River is the second longest in Africa. During the fifteenth century, Europeans called it the Zaire, a corruption of a Bantu word meaning "river." Later, the name was changed to Congo after the great African kingdom of Kongo located near its mouth. It is now known by both names.

The Congo's headwaters are in Kabalo, in the Democratic Republic of the Congo (DRC). It travels north across the country of Zaire, turns west, and then angles southwest, forming the border between Zaire and Congo. Its mouth, where it empties into the Atlantic Ocean, is in Angola.

The Congo's discharge basin is the second largest in the world. It draws water from tributaries in the Central African Republic, Cameroon, Congo, Angola, Zaire, and northern Zambia. Some of its tributaries include the Chambezi, Lualaba, Luvua, Lukuga, Chenal, Ubangi, Sangha, and Kwa. Along its course it forms many lakes, such as Bangweulu, Mweru, and Malebo Pool. Rapids, waterfalls, gorges, and a narrow channel, are found along its length.

The river straddles the equator, passing through the great rain forest of Zaire, which keeps it supplied with rainfall all year long. The Congo basin has uniform average annual rainfall of 90 inches (229 centimeters).

Annual volume of discharge of the Congo River is second only to that of the Amazon. Its force is so great that brownish-colored Congo waters can still be distinguished for more than 300 miles (480 kilometers) out to sea. During prehistoric times its flow cut a submarine canyon as deep as 4,000 feet (1,220 meters) into what is now the ocean floor.

Congo (Zaire) River
Location: Zambia, Zaire, Congo, and Angola
Length: 2,716 miles (4,346 kilometers)
Discharge Basin: 1,425,000 square miles (3,705,000 square kilometers)

In the northern and southern parts of its basin it passes through grasslands (savannas) where trees are scattered. In Zaire and along the Zaire-Congo border, dense rain forest occupies the basin. The Congo River basin contains the second largest rain forest in the world, second to the Amazon rain forest.

Vegetation in the central basin is abundant and diverse. Huge rain forest trees draped with clinging vines line its banks. In the southern regions, acacias are common. In the highlands, bamboo, heather, and giant senecio are found. Cultivation and burning have altered the vegetation along much of the Congo's length, and many non-native plants, such as cassava, citrus, and maize have been introduced.

At least 686 species of fish have been identified in Congo waters, 80 percent of which are found nowhere else in the world. Among the best known is the lungfish, which can live in oxygen-depleted water. Mammals typical of grasslands, such as lions and zebras, are found in the northern and southern regions, while gorillas and rhinoceroses are found in the rain forests. Elephants and leopards are common throughout.

The first people to inhabit the Congo were the Pygmy. Most native peoples who live along the Congo catch its fish. Some are also farmers. Many peoples, such as the Bobangi and the Teke, use the river for trade. The river is important as a source of hydroelectric power but the area's economy has not yet been able to take full advantage of its potential.

In the late nineteenth century, many European explorers, such as David Livingstone (1813–1873) and Henry Morton Stanley (1841–1904), penetrated the African interior by traveling on the Congo.

The Tigris and Euphrates   The Tigris (TY-grihs) river flows though southeastern Turkey and Iraq. Above Basra in southern Iraq, it joins the Euphrates (yoo-FRAY-teez) to form the Shatt-al-Arab, which empties into the Persian Gulf.

The Tigris-Euphrates basin is a hot, dry plain with daytime temperatures as high as 120°F (49°C). Average annual rainfall is only 8 inches (20 centimeters), and it falls seasonally, from November to March. The Euphrates floods twice during the year, the largest occurring in April and May, and creates marshy regions along its banks.

Desert vegetation, such as thorn and scrub, predominate. Bats, jackels, and wildcats are common mammals, but reptiles are more numerous, particularly snakes and lizards. Many fish live in the river, including the Tigris salmon.

Tigris and Euphrates River
Location: Turkey, Syria, and Iraq
Length: Tigris 1,181 miles (1,890 kilometers); Euphrates 1,740 miles (2,784 kilometers)
Discharge Basin: Tigris 145,000 square miles (377,000 square kilometers); Euphrates 295,000 square miles (767,000 square kilometers)

Both rivers are used for irrigation, and two major reservoirs are located on the Euphrates where farmers raise cereal grains and dates and the nomadic tribes raise livestock. The Tigris provides irrigation waters for the cultivation of wheat, barley, millet, and rice.

The Tigris-Euphrates basin was the birthplace of the great Sumerian civilization, which dates back to at least 3000 BC in what was then Mesopotamia. The Sumerians dug canals along the Tigris, and Nineveh, the capital of ancient Assyria, was located there.

The Mississippi The Mississippi is the second longest river in the United States, and its name in the Algonquian language means "father of waters." Its headwaters are in Minnesota, and its drainage basin includes at least part of thirty-one states and covers about 40 percent of the country. As it flows south, it forms the boundary between Minnesota, Iowa, Missouri, Arkansas, and Louisiana on the west and Wisconsin, Illinois, Kentucky, Tennessee, and Mississippi on the east.

Fed by Lake Itasca, a glacial lake, the upper Mississippi forms rapids where it joins the Minnesota River. Further south it is lined by high bluffs, and below Cairo, Illinois, it becomes an old stream, meandering through flat floodplains between natural levees.

Its most important tributaries are the Illinois, the Chippewa, the Black, the Wisconsin, the Saint Croix, the Iowa, the Des Moines, the Ohio, the Arkansas, the Red, and the Missouri Rivers. The mouth of the Mississippi is at the Gulf of Mexico where its delta covers a region 9,872 square miles (25,568 square kilometers) in area, and about 79,935,000 tons (72,515,812 metric tons) of sediments are deposited each year.

The river travels the length of the United States and precipitation varies from 40 to 60 inches (102 to 152 centimeters). The southern section of the river is prone to severe flooding, and in 1927 Cairo, Illinois, remained flooded for 153 consecutive days. Artificial levees, meander cutoffs, flood outlets, and upstream reservoirs have been built since that time to try to contain floodwaters. The river flooded again in 1993, covering 14,000,000 acres (5,500,000 hectares) and causing fifty deaths.

Willow, oak, and pine trees are common, as are many species of grasses and aquatic plants. Fish include catfish and paddlefish. Cranes, ducks, and other waterfowl frequent the river, as do turtles, muskrats, and deer.

Ocean-going ships can travel upstream to Baton Rouge, Louisiana. Smaller ships can travel through 15,000 miles (24,150 kilometers) of the

Mississippi River
Location: Minnesota, Iowa, Missouri, Arkansas, Louisiana, Wisconsin, Illinois, Kentucky, Tennessee, and Mississippi
Length: 2,348 miles (3,757 kilometers)
Discharge Basin: 1,243,700 square miles (3,233,620 square kilometers)

entire river system, and use of the river for transportation is increasing. Cargo consists mainly of midwestern grain and petroleum products from the Gulf of Mexico. More than 460,000,000 tons (420,000,000 metric tons) of freight are transported on the Mississippi each year.

Native peoples who once lived along the river were the Ojibwa, the Winnebago, the Fox, the Sauk, the Choctaw, the Chickasaw, the Natchez, and the Alabama. The first European to reach the river inland was Hernando de Soto (c. 1496–1542) in 1541. In 1673, Jacques Marquette (1637–1675) and Louis Jolliet (1645–1700) explored the northern regions. The river system enabled Europeans to settle the central United States, and it was an important transportation route during the Civil War.

The Missouri   The Missouri is the longest river in the United States and the largest tributary of the Mississippi. Its headwaters are the Jefferson, Gallatin, and Madison Rivers in Montana, and it travels through North and South Dakota to finally join the Mississippi in St. Louis, Missouri. Its own tributaries include the Little Missouri, the Cheyenne, the White, the Niobrara, the James, the Big Sioux, the Platte, and the Kansas rivers. Its drainage basin includes parts of Canada.

In Montana and North and South Dakota, it passes through high plains where the soil is poor. Farther south it enters the humid grain belt and the drier grasslands where cattle graze. The soil is rich along is southern stretches, although erosion is heavy. The river carries so much sediment that is nicknamed is "Big Muddy."

Precipation ranges from 20 to 40 inches (51 to 102 centimeters) annually. The river's average discharge is about 64,000 cubic feet (1,810 cubic meters) per second. Flood season is from April to June, and the federal government has built dams and reservoirs to help control flooding, provide irrigation, and make the river navigable.

Natural vegetation in the Missouri basin consists primarily of grasses. Fish include bass and trout. Mammals such as deer and coyotes can be found in this region.

Native peoples who lived in the region include the Cheyenne, the Crow, the Mandan, the Pawnee, and the Sioux. French explores Jacques Marquette (1637–1675) and Louis Jolliet (1645–1700) explored the region by 1673. By 1843 farmers had begun to settle in the Missouri valley. Today, the Missouri is important to traffic in bulk freight such as grain, coal, steel, petroleum, and cement.

Missouri River
Location: Montana, North Dakota, South Dakota, Nebraska, Iowa, and Missouri
Length: 2,466 miles (3,946 kilometers)
Discharge Basin: 529,000 square miles (1,375,400 square kilometers)

The Volga   The Volga is the longest river in Europe. Its headwaters are in the hills north of Moscow, Russia, and it empties into the Caspian Sea, which lies on Russia's southern border. It has more than 200 tributaries, the most important of which is the Kama, Oka, Vetluga, and Sura rivers The Volga delta covers more than 34,000 square miles (86,000 square kilometers).

Melting snow and summer rains are the Volga's main source of water. Annual precipitation averages 20 inches (50 centimeters) in the north and 12 inches (30 centimeters) in the south. Spring flooding is controlled by several dams and reservoirs.

The Volga passes through forested land and forms several lakes in the north. Below that it enters a flat, swampy basin bordered by low hills. Its southern section curves around mountains and finally enters a broad floodplain.

Willow, pine, and birch are common trees. River fish include sturgeon, perch, pike, and carp. Mammals include deer, beavers, and foxes. The Volga delta, with a length of 99 miles (160 kilometers), is Europe's largest estuary and the only place in Russia where pelicans, flamingo, and lotuses may be found.

Although almost all of the Volga is navigable, ice covers most of the length of the river for three months. The river carries about 50 percent of Russia's river freight, including timber, petroleum, coal, salt, farm equipment, construction materials, fish, and fertilizers.

For More Information

BOOKS

Angelier, Eugene, and James Munnick. *Ecology of Streams and Rivers.* Enfield, New Hampshire: Science Publishers, 2003.

Day, Trevor. *Biomes of the Earth: Lakes and Rivers.* New York: Chelsea House, 2006.

Gleick, Peter H., with Nicholas L. Cain, et al. *The World's Water 2004-2005: The Biennial Report on Freshwater Sources.* Washington DC: Island Press, 2004.

Gloss, Gerry, Barbara Downes, and Andrew Boulton. *Freshwater Ecology: A Scientific Introduction.* Malden, MA: Blackwell Publishing, 2004.

Hauer, Richard, and Gary A. Lamberti. *Methods in Stream Ecology,* 2nd Edition. San Diego: Academic Press/Elsevier, 2007.

Postel, Sandra, and Brian Richter *Rivers for Life: Managing Water For People And Nature.* Washington DC: Island Press, 2003.

Volga River
Location: Russia
Length: 2,292 miles (3,667 kilometers)
Discharge Basin: 532,800 square miles (1,385,280 square kilometers)

Voshell, J. Reese, Jr. *A Guide to Common Freshwater Invertebrates of North America.* Blacksburg, VA: McDonald and Woodward Publishing Company, 2002.

Woodward, Susan L. *Biomes of Earth: Terrestrial, Aquatic, and Human Dominated.* Westport, CT: Greenwood Press, 2003.

Worldwatch Institute, ed. *Vital Signs 2003: The Trends That Are Shaping Our Future.* New York: W.W. Norton & Co., 2003.

PERIODICALS

Carwardine, Mark. "So Long, and Thanks for All the Fish: If the Yangtze River Dolphin isn't Quite Extinct Yet, It Soon Will Be." *New Scientist.* 195. 2621 September 15, 2007: 50.

Petty, Megan E. "The Colorado River's Dry Past." *Weatherwise.* 60. 4 (July-August 2007): 11.

Sterling, Eleanor J, and Merry D. Camhi. "Sold Down the River: Dried Up, Dammed, Polluted, Overfished—Freshwater Habitats Around the World are Becoming Less and Less Hospitable to Wildlife." *Natural History.* 116. 9 November 2007: 40.

ORGANIZATIONS

American Rivers, 1101 14th Street NW, Suite 1400, Washington, DC 20005, Phone: 202-347-7550; Fax: 202-347-9240, Internet: http://www.americanrivers.org.

Environmental Defense Fund, 257 Park Ave. South, New York, NY 10010, Phone: 212-505-2100; Fax: 212-505-2375; Internet: http://www.edf.org.

Envirolink, P.O. Box 8102, Pittsburgh, PA 15217; Internet: http://www.envirolink.org.

Environmental Protection Agency, 401 M Street, SW, Washington, DC 20460, Phone: 202-260-2090; Internet: http://www.epa.gov.

The Freshwater Society, 2500 Shadywood Rd., Navarre, MN 55331, Phone: 952-471-9773; Fax: 952-471-7685; Internet: http://www.freshwater.org.

Friends of the Earth, 1717 Massachusetts Ave. NW, 300, Washington, DC 20036-2002, Phone: 877-843-8687; Fax: 202-783-0444; Internet: http://www.foe.org.

Greenpeace USA, 702 H Street NW, Washington, DC 20001, Phone: 202-462-1177; Internet: http://www.greenpeace.org.

Global Rivers Environmental Education Network (GREEN), 2120 W. 33rd Avenue, Denver, CO 80211; Internet: http://www.earthforce.org/green.

Izaak Walton League of America, 707 Conservation Lane, Gathersburg, MD 20878, Phone: 301-548-0150; Internet: http://www.iwla.org/.

Project Wet, 1001 West Oak, Suite 210, Bozeman, MT 59717, Phone: 866-337-5486; Fax: 406-522-0394; Internet: http://projectwet.org.

Sierra Club, 85 2nd Street, 2nd Fl., San Francisco, CA 94105, Phone: 415-977-5500; Fax: 415-977-5799, Internet: http://www.sierraclub.org.

World Wildlife Fund, 1250 24th Street NW, Washington, DC 20090-7180, Phone: 202-293-4800; Internet: http://www.wwf.org.

WEB SITES

National Geographic Magazine: http://www.nationalgeographic.com (accessed September 12, 2007).

National Park Service: http://www.nps.gov

Nature Conservancy: http://www.tnc.org

Scientific American Magazine: http://www.sciam.com

U.S. Fish and Wildlife Services: http://www.fws.gov/

Seashore

The seashore, also called the coastline, shoreline, or beach, is the portion of a continent or island where the land and sea meet. The seashore includes the area covered by water during high tide and exposed to air during low tide, the area splashed by waves but never under water, and the area just beyond the shore that is always under water. (The tides are the rhythmic rising and falling of the sea.) Seashores vary greatly in appearance, from flat, sandy, and washed by gentle waves to storm-battered and rocky or bounded by tall cliffs.

How the Seashore Is Formed

Seashores were first created when the continents and islands of Earth were formed. Since then, many changes have occurred. Some happened during prehistoric times and others are still taking place.

Movement of Earth's crust   The underlying structure of the shoreline depends upon the shape of the land where it meets the ocean and the type of rock of which it is a part. Earthquakes and volcanoes during prehistoric times may have helped form many shorelines. An earthquake, for example, caused part of the California shoreline to sink. The sunken area became what is now San Francisco Bay and a new shoreline was created. In regions like northern California, where earthquakes and volcanoes still occur, the shoreline may undergo many more changes in the future.

Glaciers   During prehistoric times, glaciers (giant, slow-moving rivers of ice) may have altered the shape of the seashore by cutting into it and leaving deep valleys behind when they retreated. Glaciers created the fjords (fee-OHRDS; long, narrow arms of the ocean stretching inland) of Scandinavia, Greenland, Alaska, British Columbia, and New Zealand. They also created the U-shaped valleys along the coastline of southern

WORDS TO KNOW

Estuary: The place where a river traveling through lowlands meets the ocean in a semi-enclosed area.

Intertidal zone: The seashore zone covered with water during high tide and dry during low tide; also called the mitermle, or the littoral, zone.

Littoral zone: The area along the shoreline that is exposed to the air during low tide; also called intertidal zone.

Longshore currents: Currents that move along a shoreline.

Neap tides: High tides that are lower and low tides that are higher than normal when the Earth, sun, and moon form a right angle.

Rip currents: Strong, dangerous currents caused when normal currents moving toward shore are deflected away from it through a narrow channel; also called riptides.

Spring tides: High tides that are higher and low tides that are lower than normal because the Earth, sun, and moon are in line with one another.

Sublittoral zone: The seashore's lower zone, which is underwater at all times, even during low tide.

Submergent plant: A plant that grows entirely beneath the water.

Supralittoral zone: The seashore's upper zone, which is never underwater, although it may be frequently sprayed by breaking waves; also called the splash zone.

Tidal bore: A surge of ocean water caused when ridges of sand direct the ocean's flow into a narrow river channel, sometimes as a single wave.

Tsunami: A huge wave or upwelling of water caused by undersea earthquakes that grows to great heights as it approaches shore.

Glaciers continue to alter the shape of shores by cutting into them, and leaving valleys behind when they retreat. COPYRIGHT © 2003 KELLY A. QUIN.

Chile. In polar regions, glaciers are still at work, carving deep channels as they inch toward the sea.

The presence of a glacier weighs the land down, causing it to sink. About 10,000 years ago when many glaciers receded, some coastlines rose up, and areas once under water were now above it. In some places, such as Scandinavia, the coastlines are rising as much as 0.4 inch (10 millimeters) a year even though the glaciers have been gone for thousands of years.

Changes in sea level Sea level refers to the average height of the sea when it is halfway between high and low tides. Sea level changes over time. For example, when the Ice Ages ended and glaciers began to melt into the ocean, the

Locked in the Ice

Exploration of remote areas, such as Antarctica, demanded people with courage and endurance. Ernest Shackleton (1874–1922), a British explorer, had those characteristics and made several expeditions to Antarctica in the early 1900s. His most remarkable, and almost fatal, journey was made in 1914 with twenty-seven other men on board a small ship called *The Endurance*, a name that would prove prophetic.

The Endurance became trapped in drifting ice in the Weddell Sea (part of the Antarctic Ocean) during one of the coldest winters in memory. The explorers had come prepared to spend the winter, if necessary, but the weather did not get warm enough the following spring to free the ship. The

ice crushed its forward section, and the ship sank in November 1915. By then the men had begun to kill penguins and their own sled dogs for food.

They managed to save three small boats and, for seven days, used them to travel to an uninhabited island. All were starved and exhausted, but taking no time to rest on the island, Shackleton and five of the strongest men set out again in one of the boats to seek aid. They finally found it on South Georgia Island farther north. Shackleton turned around and went back to rescue the others. The first three tries were unsuccessful, but he succeeded on the fourth attempt. Everyone survived because of Shackleton's courage and endurance.

level of the water rose. Some areas that were once exposed to the air were now covered permanently by water and new shorelines were created.

Sea level is still changing. For the past 100 years, it has risen about 0.08 inch (2 millimeters) per year. As the water creeps higher, more of the existing shoreline becomes submerged. Low-lying coastal areas in Texas and Louisiana have already been flooded.

Offshore barriers   The presence of a barrier island, reef (an underwater wall made of rocks, sand, or coral), or other offshore landmass running parallel to the shore may affect a seashore's formation. It does this by reducing the effects of wind and water. Deposits of sediment (particles of matter) are allowed to accumulate because the force of the waves is lessened.

Coral reefs are created in tropical regions by small, soft, jellylike animals called corals and algae that trap hard calcium carbonate. Corals attach themselves to hard surfaces and build a shell-like external skeleton. Many corals live together in colonies, the younger building their skeletons next to or on top of older skeletons. Gradually, over hundreds, thousands, or millions of years, a reef of these skeletons is formed. Because reefs slow the movement of the water, sediments sometimes lodge in the reef allowing plants to take root. As the plants die and decay, a layer of soil is created. Eventually, the shoreline may extend, and even trees may grow.

SEASHORE PARKS AND RESERVES OF THE WORLD (PARTIAL LIST)

Name	Location	Square miles (square kilometers)	Features
Aldabra Islands Nature Reserve	Seychelles (Africa)	60 (156)	Protects giant tortoise and other animals
Bako National Park	Sarawak	10 (26)	Bays and caves; forest; wild pigs, deer, monkeys, birds
Cape Le Grand National Park	Australia	124 (321)	Beach; coostal plants
Easter Island National Park (Rapa Nui)	Chile	63 (163)	Protects vegetation and animals
Elat Gulf Coral Reef Reserve	Israel	0.5 (1.3)	Coral reefs with tropical fish
Eldey Nature Reserve	Iceland	0.006 (0.015)	Gannet breeding area
Franklin D. Roosevelt National Park	Uruguay	58 (150.8)	Sand dunes, pines
Ise-Shima	Japan	201 (522.6)	Forested coastline; pearl farms
Kong Karls Land Reserve	Norway	200 (520)	Arctic islands, polar bears
Kranji Reserve	Singapore	0.08 (0.21)	Mangrove marsh
Kyzylagach	Azerbaijan	363 (943.8)	Coastal reed and salt marshes; flamingos, bustards
Pembrokeshire Coast	Great Britain	225 (585)	Rocky coast; island birds
Prince Edward Island	Canada (Gulf of St. Lawrence)	7 (18.2)	Forested coastline; many small mammals
Rikuchu Kaigan	Japan	45 (117)	Cliffs, islands, and beaches; forests; birds
Skallingen	Denmark	12 (31.2)	Dunes, marshes
St. Lucia	Natal (Africa)	190 (494)	Estuary on False Bay; hippopatoami and birds
Westhoek Nature Reserve	Belgium	1.3 (3.4)	Sand dunes, marine plants
Yala National Park	Sri Lanka	91 (236.6)	Lagoons, rocky hills

The Bordering Sea

The oceans are constantly, restlessly moving. This movement takes place in the water column (the water in the ocean exclusive of the sea bed or other landforms) in the form of tides, waves, and currents; all of which affect the shoreline.

Tides   Tides are rhythmic movements of the oceans that cause a change in the surface level of the water. When the water level rises, it is called high tide. When it falls, it is called low tide. Along some shorelines, the tides are barely measurable. In other areas the difference between high and low tide may be several feet (meters). High and low tides occur in a particular place at least once during each period of 24 hours and 51 minutes.

Tides are caused by a combination of the gravitational pull of the sun and moon and Earth's rotation. The sun or moon pulls on the water, causing it to bulge outward. At the same time, the centrifugal force (movement from the center) created by Earth's rotation causes another bulge on the opposite side of Earth. Both of these areas experience high tides. At the same time, water is pulled from the areas in between, and those areas experience low tides.

When the Earth, sun, and moon are lined up, the gravitational pull is strongest. At these times, high tides are higher and low tides are lower than normal. These are called spring tides. When the Earth, sun, and moon form a right angle and the gravitational pull is weakest, high tides are lower and low tides are higher than normal. These are called neap tides.

Tidal bores are surges of tidal waters caused when ridges of sand block the flow of ocean water and direct it into a narrow channel, sometimes as a single wave. Most tidal bores are harmless, but the bore that enters the Tsientang River in China sometimes reaches 15 feet (4.5 meters) in height and travels 25 feet (7.5 meters) per second.

Waves   Waves are rhythmic rising and falling movements in the water. Most surface waves are caused by wind. Their size is due to the speed of the wind, the length of time it has been blowing, and the distance over which it has traveled. A breaker is a wave that collapses on a shoreline in a mass of foam. As it rolls in from the ocean, the bottom of the wave is slowed by friction as it drags along the sea floor. The top then outruns it and topples over, landing on the shore.

The Beaches at Normandy

Beaches have often been used by invading armies to conquer the neighboring region. During World War II (1939–45), the beaches of the province of Normandy, France, were the site of an important strike against Nazi Germany. Germany had conquered France earlier in the war and much of their power in that country was concentrated in Normandy, which bordered the English channel and was only a short distance from England, which the Nazis hoped to conquer.

On D-Day, June 6, 1944, in an effort to recapture France, the Allies (Britain, Canada, Poland, France, and the United States) struck the beaches at Normandy. The beaches, named Utah, Omaha, Gold, Sword, and Juno, stretch about 50 miles (80 kilometers) along the coast between Cherbourg and LeHavre. The slope of the beaches is gentle and a wide expanse of shore is exposed at low tide. The date for the invasion was selected based on the tides, the weather, the presence of moonlight, and other conditions.

The Germans, who expected trouble, were unaware of the exact invasion point and had fifty infantry and ten tank divisions spread out over France and neighboring countries. To distract the Germans, British-based aircraft bombed rail lines, bridges and airfields on French soil for two months before D-Day. The night before, paratroops were dropped inland to interfere with enemy communications. Naval guns pounded German gun nests on shore.

In the early daylight at low tide on June 6, in rough seas, about 5,000 Allied ships approached the Normandy coastline. The British and Canadians moved in efficiently at Gold, Juno, and Sword beaches and the Americans at Utah Beach. But at Omaha Beach, the central point of the landings, American troops encountered heavy German gunfire, and many men were killed. Within five days, sixteen Allied divisions had landed in Normandy. By August, Paris had been freed, and it was the beginning of the end of Nazi rule in Europe.

A realistic account of the Normandy invasion can be viewed in Stephen Spielberg's 1998 Academy-Award-winning film *Saving Private Ryan*.

Waves can be extremely powerful. Storm waves can even hurl large rocks high into the air. So many rocks have broken the beacon at the lighthouse at Tillamook Rock, Oregon, that the beacon—133 feet (40.5 meters) above the water—is now enclosed in a steel grating.

One dangerous type of wave, called a tsunami (soo-NAH-mee), is caused mainly by undersea earthquakes. When the ocean floor moves during the quake, its vibrations create a powerful wave that travels to the surface. When tsunamis strike coastal areas, they can destroy entire towns and kill many people.

Currents   Currents are the flow of the water in a certain direction. Most currents are caused by the wind, the rotation of Earth, and the position of continental landmasses. A longshore current is one that runs along a shoreline. Rip currents, or riptides, are strong, dangerous currents that occur when currents moving toward shore are deflected away from it through a narrow channel.

Upward and downward movement also occurs in the ocean. Vertical currents are primarily the result of differences in the temperature and salinity (level of salts) of the water. In some coastal areas, strong wind-driven currents carry warm surface water away. Then an upwelling (rising) of cold water from the deep ocean occurs to fill the space. This is more common along the western sides of continents. These upwellings bring many nutrients from the ocean floor to the surface, encouraging a wide variety of marine (ocean) life.

Zones in the Seashore

The seashore can be divided into zones based on relationships to the ocean, particularly the tides.

The seashore's lowest zone is underwater at all times, even during low tide. This lower zone is called the sublittoral zone. It is a marine environment.

The intertidal zone (called the middle, or littoral, zone) is covered during high tide and exposed during low tide. This is the harshest of all seashore environments, since any animal or plant that lives here must be able to tolerate being submerged for part of the day and exposed to the air and sun for the rest.

The upper zone is never underwater and may be frequently sprayed by breaking waves. It is often referred to as the splash or supralittoral zone. Conditions here are similar to other dry-land areas.

Climate

The climate of a particular seashore depends upon its location. In general, if the shoreline is part of a desert country, such as Saudi Arabia, the climate will be hot and dry. If it is part of a country in the frozen north,

The Silver Dragon

High tides arrive at most shorelines gradually. In locations where tapering coastlines face a large ocean, however, the volume of the tide is focused very narrowly and its force increases. When such a tide meets a river flowing into the sea, a steep wave forms, forcing itself up the river and far inland. This wave is called a tidal bore.

The world's largest tidal bore is called the Silver Dragon, and it occurs once a month where the South China Sea invades the Quiantang River of China. Reaching heights of 20 feet (6 meters) or more, the Silver Dragon rushes over an area 5 miles (8 kilometers) wide.

Other large bores occur on the Amazon River in South America, on the Petitcodiac where it meets the Bay of Fundy in Canada, on the Hooghly in India where it flows into the Bay of Bengal, and on the Severn in England.

Hurricane winds originate at sea, but their forceful rotating winds cause extensive damage, both from the winds of up to 180 miles per hour (300 kilometers per hour), and the huge waves that batter the shoreline. IMAGE COPYRIGHT KONOVALIKOV ANDREY, 2007. USED UNDER LICENSE FROM SHUTTERSTOCK.COM.

such as Siberia, the climate will be cold. The presence of the ocean can create some climatic differences.

The ocean absorbs and retains some of the sun's heat. In the winter, the warmer water releases heat into the colder atmosphere, helping to keep temperatures warmer inland. In summer, when the water temperature is cooler than the air temperature, winds off the ocean help cool coastal areas.

Other effects of the ocean on climate result primarily from moderating ocean currents and storms at sea.

Moderating currents   Ocean currents may be warm or cold, depending upon where they started. The presence of a current can moderate the weather along a particular coastline. For example, the North Atlantic Drift is a warm current that originates in regions farther south. As it flows around the coast of Scotland, it warms the air temperature enough that palm trees will grow there. As it travels farther north along the coast of Norway, it keeps the water above freezing so that Norway's ports are open all winter, even though they are above the Arctic Circle.

Storms at sea   Seashores are vulnerable to storms that originate at sea and produce strong winds and high waves. Hurricanes and typhoons are violent tropical storms that form over the oceans. Their wind speeds can reach 75 to 200 miles (120 to 322 kilometers) per hour. Their forceful, rotating winds cause much damage when they reach land, as do the waves that batter the shoreline.

The worst coastal weather in the world is in the North Atlantic where the climate is cooler and the force of the waves has been measured at 6,000 pounds per square foot (13,200 kilograms per square meter). Driven by gale-force winds, waves along these coastlines have been known to be very destructive.

Geography of the Seashore

The geography of the seashore is affected by the process of erosion (wearing away) and deposition (dep-oh-ZIH-shun; setting down), which helps determine the different types of surface and landforms found at the seashore.

Erosion and deposition As waves crash against a shoreline, they compress (squeeze) the air trapped in cracks in rocks. As the waves retreat, the pressure is suddenly released. This process of pressure and release widens the cracks and weakens the rock, causing it to eventually break apart. Some waves, especially those created by storms, are very high and forceful. In places where wave action is strong, the waves pick up particles of rock and sand and throw them against the shoreline with a crashing motion. This produces a cutting action.

Some of the chunks and particles eroded from a shoreline may then be carried out to sea by waves. The particles sink and become deposited on the ocean floor. Other particles may be carried by longshore currents farther along the coast and deposited where there is shelter from the wind, and the wave action is not as severe. Even boulders may be carried by this means.

> ## In the Path of the Great Waves
>
> Over 2.8 million people have been killed by tsunamis (giant waves) during the past twenty years. One in every three tsunamis occurs off the coast of Chile, and that country experiences 40 percent of all the damage caused worldwide.
>
> In 1964, an earthquake struck Alaska, causing great damage and breaking a pipeline carrying oil. The oil caught fire. Then a tsunami three stories tall followed the quake and, picking up the burning oil, carried it overland in a tide of fire. The fires reached the railroad yards, where the steel tracks soon glowed red from the heat. When more tsunamis followed and flowed over the tracks, the sudden cooling made the tracks rise up and curl like snakes.

Shoreline surfaces Seashore surfaces are classified as rocky, sandy, or muddy, depending upon their composition (makeup).

Rocky shores may consist of vertical cliffs, sloping shores, platforms, and boulder-covered areas. Vertical cliffs have no protection from the waves. On shores covered by boulders, spaces among the stones are protected, and pools of water, called tide pools, may form in them.

Beaches along rocky shorelines usually consist of pebbles and larger stones. The waves carry away the finer particles.

Sandy shores make up about 75 percent of the world's seashores that are not ice-covered. They are constantly changing, depending upon the movement of the wind, the water, and the sand.

Some sandy shores form a steep slope down to the sea. On this type of shoreline, the waves break directly on the beach. Cycles of erosion and deposition are more extreme and have a greater effect on the shape of the shoreline. The greater the force of the waves, the steeper the slope, and the larger the sand particles.

Sandy shores that slope gently down to the sea are usually protected in some way from the full force of the water. A reef, for example, may have

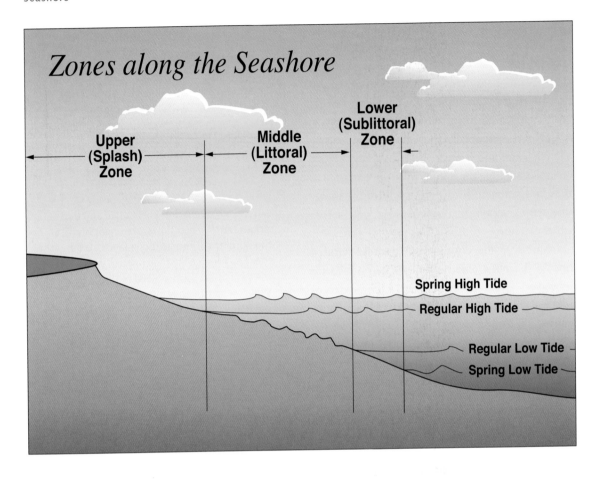

Zones along the Seashore

Upper (Splash) Zone

Middle (Littoral) Zone

Lower (Sublittoral) Zone

Spring High Tide

Regular High Tide

Regular Low Tide

Spring Low Tide

formed some distance from shore. As a result, the waves that reach the beach are gentle, erosion is slow, and sand particles are finer.

The texture of the sand helps determine what the beach will be like. Fine sand packs down and produces a smooth, gentle slope. Coarse sand allows the waves to sink in and move the particles around, producing a steep surface.

An estuary is where a river, traveling through lowlands, meets the ocean in a semi-enclosed channel or bay. In these gently sloping areas, river sediments (soil and silt) collect, and muddy shores form. Water in an estuary is brackish, a mixture of fresh and saltwater.

Estuaries are semi-enclosed by land, which usually protects them from the full force of ocean waves. High tide may bring in a new supply

Lighthouses are found on some seashores, built to guide boats into ports and docks, or away from dangerous waters. COPYRIGHT © 2006 KELLY A. QUIN.

of seawater and the water becomes more salty. During low tide, fresh water from the river dominates. After a rain washes soil into the river, the water carried into the estuary may be cloudy from the added sediments.

Landforms   Landforms found at the seashore include cliffs and rock formations; beaches and dunes; deltas; spits and bars; and certain types of saltwater wetlands.

Cliffs and rock formations   High cliffs usually occur when highlands meet the sea. Pounding waves may gradually eat into the base of the cliffs, eroding the rock and creating a hollowed-out notch. Eventually, the over-hang collapses and falls onto the beach, creating a platform of rock and soil.

Many coastlines consist of both hard and soft rock. Wave action erodes the soft rock first, sometimes sculpting strange and beautiful shapes, such as arches. Caves may be carved into the sides of cliffs, or headlands may be created. A headland is a large arm of land made of hard rock that juts out from the beach into the ocean after softer rock has been cut away.

Beaches and dunes   Beaches are almost-level stretches of land along the water's edge. They may be covered by sand or stones. Sand is small particles of rock less than 0.08 inch (2 millimeters) in diameter. It may be white, golden, brown, or black, depending upon the color of the original rock. Yellow sand usually comes from quartz and black sand from volcanic

Once grasses take root in dunes, the dunes are less likely to travel with the wind and change their position and shape. COPYRIGHT © 2006 KELLY A. QUIN.

rock. White or pink sand may have been formed from limestone, seashells, or coral particles.

Sand is carried not only by water but also by wind. When enough sand has been heaped up to create a ridge or hill, it is called a dune. Dunes range from 15 to 40 feet (4.6 to 12.2 meters) in height, and a few may become 75 feet (23 meters) tall. Individual dunes often travel, as the wind changes their position and shape. They tend to shift less if grasses take root in them and help hold them in place.

Deltas   In estuaries, where rivers meet the sea, huge amounts of silt (very small particles of soil) carried from far inland by the river are deposited in the ocean along the shoreline. Large rivers can dump so much silt that islands of mud build up, forming a fan-shaped area called a delta. In regions where Earth's crust is thin, the weight of these sediments may cause the shoreline to sink. Huge deltas have formed at the mouths (where a river empties into a larger river or ocean) of the Mississippi River in the United States and the Nile River in Egypt.

Spits and bars   A spit is a long, narrow point of deposited sand, mud, or gravel that extends into the ocean. In an estuary, a spit may reach up into the river's mouth.

A bar is an underwater ridge of sand or gravel formed by tides or currents that extends across the mouth of a bay (an area of the ocean partly enclosed by land). If the bar closes off the bay completely, the water trapped behind the bar is called a lagoon. A tombolo is a bar that forms between the beach and an island, linking them together.

Saltwater wetlands Saltwater wetlands are portions of land covered or soaked by ocean water often and long enough to support plants adapted for life under those conditions.

A swamp is a type of wetland dominated by trees. A saltwater swamp is formed by the movement of the ocean tides. When the tide is low, flat places in the swamp are not under water.

Marshes are wetlands dominated by non-woody plants such as grasses, reeds, rushes, and sedges. Saltwater marshes are found in low, flat, poorly drained coastal areas. They are especially common in deltas, along low-lying seacoasts, and in estuaries. Saltwater marshes are greatly affected by the tides, which raise or lower the water level on a daily basis. A saltwater marsh may have tidal creeks, tidal pools, and mud flats.

The Legendary Lusca

Along the shores of the Bahama Islands, in the depths of the beautiful blue waters of the Caribbean Sea, lives the legendary Lusca, a creature that is supposedly half shark and half octopus. The Lusca draws people and even boats into its lair and, afterwards, like a rather rude dinner companion, it signals its satisfaction with a sudden upwelling "burp" of water.

Although not really a sea monster, the Lusca can be a bit tricky to get to know. A Lusca is not a creature at all, but an underwater cave carved out of the limestone that lies under the Bahama Islands. As sea levels rise and fall, these caves fill with water. When rainwater mixes with the seawater, their different densities (weights) cause whirlpools. These whirlpools account for people being drawn down into the caves and for the sudden burps of water. These extreme conditions often churn up sediments, making the caves dangerous for divers who, in the resulting dimness, may not be able to find their way out again.

Plant Life

Seashore plants include many kinds of seagrasses and some species of trees. The types of plants that grow in a particular region are determined by climate and the kind of shoreline surface—rocky, sandy, or muddy. Some sandy or muddy shores do not support plants because the soil is frequently disturbed.

Plants that live in the lower (sublittoral) zone along the seashore are always surrounded by water. Most have not developed the special tissues and organs for conserving water that are needed by plants on land. The surrounding water offers support to these plants, helping to hold them upright. Their stems are soft and flexible, allowing them to move with the currents without breaking.

Plants that live in the intertidal (littoral) zone are very hardy because they are exposed to the air for part of the day and are underwater for part of the day. Plants in the upper (supralittoral), or splash, zone have adapted to life on land. However, they must be tolerant of salty conditions.

The Web of Life: Seashore Succession

Life exists everywhere on our planet, even in the most remote, unfriendly places. Often, one life form leads the way for others because its presence in the habitat causes changes. Soon, other life forms move in, some to eat the first inhabitants, others because the habitat is now more comfortable for them. This process of succession occurs regularly along the seashore in some surprising ways.

Suppose a new boat is tied to the end of a pier. Within minutes, life forms will have begun to build colonies on its hull. Within an hour, bacteria will attach themselves to any surface below the water line. Phytoplankton and zooplankton come next, usually within the first day. By the second or third day, hydroids and bryozoa, tiny but more complex animals, move in too. If the boat is not moved, barnacles and larger algae will have attached themselves to its hull by the end of the week and the other animals will have been greatly reduced in number. Eventually, mussels will move in, crowding everybody else out.

Algae, fungi, and lichens Most marine plants are algae, and many algae are green plants. However, it is generally recognized that algae do not fit neatly into the plant category. Most algae have the ability to make their own food by means of photosynthesis (foh-toh-SIHN-thih-sihs), the process by which plants use the energy from sunlight to change water and carbon dioxide into the sugars and starches they use for food. Algae require other nutrients they obtain from the water. In certain regions, upwelling of deep ocean waters during different seasons brings more of these nutrients to the surface. This results in sudden increases in the numbers of algae. Increases also occur when nutrients are added to a body of water by sewage, or by runoffs from fertilized farmland.

Some forms of algae, called phytoplankton (fy-toh-PLANK-tuhn), are so tiny they cannot be seen without the help of a microscope. They float freely in the water, allowing it to carry them from place to place. Other larger species, often referred to as seaweeds, may be anchored to the seafloor.

Fungi (FUHN-ji) cannot make their own food by means of photosynthesis. Some, like molds, obtain nutrients from dead or decaying organic matter. They assist in the decomposition (breaking down) of this matter and release many nutrients needed by other plants. Others are parasites and attach themselves to, and feed on, other living organisms.

Lichens are combinations of algae and fungi that live in cooperation; the fungi surround the algal cells, and the algae obtain food for themselves and the fungi by means of photosynthesis. Lichens prefer the upper zones of the seashore, surviving dryness by soaking up water during high tide.

Common algae Two types of algae commonly found along the seashore include phytoplankton and seaweeds. Diatoms and dinoflagellates (DI-noh-FLAJ-uh-lates) are the most common types of phytoplankton.

Seaweeds are forms of green, brown, and red algae that grow primarily along rocky seashores. Green algae prefer the upper zone where they are exposed to sunlight and fresh water from rain. Brown algae prefer the middle zone and shallow water. They absorb sunlight readily and are tough enough to endure the action of waves and tides. Red algae live in tide pools or offshore waters, as do the large forms of brown algae, called kelps.

Seaweeds are different from land plants in that they do not produce flowers, seeds, or fruits. They lack root systems because they do not need roots to draw water from the soil. Instead, they have rootlike holdfasts that anchor them to rocks. They are seldom found on sandy shorelines because there are no rocks for anchorage.

The type of shoreline determines the species of seaweeds that will grow there. Different species prefer different zones along the same shoreline. Some are better adapted to dry conditions than others. Some species, for example, have a thick layer of slime that prevents water loss, while others have a layer of tissue that retains water.

Growing season Algae contain chlorophyll, a green pigment used to turn energy from the Sun into food. As long as light is available, algae can grow. In some species, the green color of the chlorophyll is masked by orange-colored compounds, giving the algae a red or brown color.

The growth of ocean plants is often seasonal. In northern regions, the most growth occurs during the summer. In temperate (moderate) zones, growth peaks in the spring but continues throughout the summer. In regions near the equator, growth is steady throughout the year.

Reproduction Algae may reproduce in one of three ways. Some split into two or more parts; each part becoming a new, separate plant. Others form spores (single cells that have the ability to grow into a new organism). A few reproduce sexually, during which cells from two different plants unite to form a new plant.

Fungi and lichens usually reproduce by means of spores. Lichens can reproduce when soredia (algal cells surrounded by a few strands of fungus) break off and form new lichens wherever they land.

Green plants   Green plants have roots, stems, leaves, and often flowers. Most green plants need several basic things to grow: light, air, water, warmth, and nutrients. Like algae, green plants depend upon photosynthesis for food. The nutrients, primarily nitrogen, phosphorus, and potassium, are obtained from the soil. These minerals may not be in large supply. In some seashores along dry, desert coasts, plants have evolved that can absorb mist from the nearby ocean through their leaves. The mist provides enough water for them to survive.

Green plants grow in all seashore zones. Those that grow in the lower zone are sea plants, and those that grow in the upper zone are land plants.

Common green plants   Green plants found along the seashore include seagrasses, such as eelgrass, turtlegrass, and paddleweed. These plants live in the lower zone but are similar to land grasses. They have roots, and they bloom underwater. Beds of seagrass occur in sandy or muddy shores in calm areas protected from currents, such as in lagoons or behind reefs. They attract a wide variety of grazing marine animals. These grasses help slow the movement of the water and prevent erosion of the shoreline.

The high salt content of the seawater in the intertidal zone makes it hard for land plants to adapt. Grasses such as marsh grass, cord grass, and salt hay grass thrive here. In salt marshes, glasswort and sea lavender are found.

Sand dunes are too dry to support much plant life, but there are many species of plants that grow on dunes. These include sandwort, beach pea, marram grass, yellow horn poppy, beach morning glory, and sea oats. Sea oats grow a long taproot (main root) that may extend more than 6 feet (1.8 meters) into the sand to reach water. Dunes may support trees, such as pine and fir, if the soil is stable and freshwater is available.

Mangrove trees grow in estuaries in tropical zones where the shoreline is muddy. Some are small, shrublike plants; others are tall and produce large forests. Mangroves have two root systems. One is used to anchor them in the muddy bottom. The other is exposed to the air from which it pulls oxygen because the soil in mangrove swamps is usually low in oxygen.

Growing season   The growing season depends upon where the seashore is located. Near the equator, growth continues throughout the year. In northern regions, there is a spurt of growth during the summer. In moderate climates, growth begins in the spring and continues throughout the summer.

Reproduction   Green plants often depend upon the wind and insects to carry pollen from the male part of a flower, called a stamen, to the female part of a flower, called a pistil, for reproduction. This process is called pollination. Others send out stems from which new plants sprout.

Endangered species   Coastlines are popular places for people to live, but plants can suffer when their natural habitat is disturbed. Algae and seagrasses can be destroyed by polluted water. Dune grasses are easily trampled by beachgoers or destroyed by dune buggies and other off-road vehicles. When visitors pick flowering plants, they limit the plants' ability to reproduce. Many mangrove forests are being reduced due to erosion of the deltas.

Animal Life

The kinds of animals that live along a particular shoreline are determined by zone (upper, intertidal, or lower) and the type of surface (rocky, sandy, or muddy). This is true of different species of the same animal. Hermit crabs, for example, live comfortably in the lower and intertidal zones of rocky shores, whereas mudcrabs prefer muddy estuaries.

> ## Bursting at the Seams
>
> Crabs, like other crustaceans, have a hard outer shell. This shell does not grow as the crab grows. Any crab that outgrows its current shell has to get rid of it fast or feel the pinch. This shedding of the shell is called molting.
>
> Molting is caused by hormones, chemical messengers in the crab's body that help split the shell so the crab can climb out. While it runs around shell-less, the crab is in greater danger from predators. Eventually, its skin hardens and soon it has a new, ready-made suit of armor.

Lower zone animals, including sea anemones, shrimp, and small fish, are underwater almost all of the time. Upper zone animals, such as ghost crabs, prefer dry conditions and live on land. The middle, or intertidal, zone along rocky shores supports the most life-forms, in spite of its being the harshest zone of all. Cockles, barnacles, clams, sea urchins, and fish are all found in this zone. Some, such as clams, survive during low tide when they are exposed to air by closing their shells to keep from drying out. Their shells also help them survive battering by the waves. Others find shelter in small pools where water collects in depressions and among rocks when the tide goes out, remaining there until high tide returns.

On depositional (sandy or muddy) shores, animals are constantly shifted around by the motion of water because there is nothing on which to anchor themselves. Many burrowing animals, such as clams and lungworms, survive by digging into the sand. They often have siphons (tubes) through which they can draw in oxygenated water and food.

Life on eroding shores is even harder; animals must attach themselves firmly to rocks or be washed away by the battering waves. The crevices of the rocks are homes to soft-bodied animals such as sea anemones. Dark caves found on many rocky shorelines provide a place where many sea creatures may take shelter.

Microorganisms   Seashores are home to many kinds of microscopic animals. Most microorganisms are zooplankton (tiny animals that drift

A hermit crab moves on a sandy beach in Mozambique, southern Africa. They search the shallow water for food. IMAGE COPYRIGHT ECOPRINT, 2007. USED UNDER LICENSE FROM SHUTTERSTOCK.COM.

with the current) that live on the surface of the sea. They include protozoa, nematodes (worms), and the larvae or hatchlings of animals that will grow much larger in their adult form. Some zooplankton eat phytoplankton and, in turn, are preyed upon by other carnivorous (meat-eating) zooplankton.

Bacteria   Bacteria are found in every zone along the shore and provide food for microscopic animal forms. They also help decay the dead bodies of larger organisms.

Much of the water that washes up on a beach sinks downward again and the sand or gravel acts as a filter. Particles of matter suspended in the water become trapped between the grains of sand. These particles become food for the bacteria that consume them and then release other nutrients into the water.

Invertebrates   Animals without backbones are called invertebrates. Many species are found along the seashore. Crustaceans and mollusks, invertebrates that usually have hard outer shells, are well adapted to the intertidal zone because their shells help prevent them from drying out during low tide. Soft-bodied animals, such as sea anemones and small octopi, prefer rocky shores where they can hide in rock crevices and survive low tide in rock pools.

Sand dunes are home to many insects, such as wasps, ants, grasshoppers, and beetles. Some insects and spiders burrow into the sand during the heat of the day and hunt for food at night.

Common seashore invertebrates   Tiny crustaceans called sand hoppers, or beach fleas, live on sandy beaches where they can hop several feet (1 meter) in one jump even though they are only 0.6 inch (1.5 centimeters) long. They are not real fleas and do not bite humans or other animals but live on seaweed and dead matter.

On rocky shores, many species of periwinkles are common. Periwinkles are snails that come in several colors, including brown, yellow, and blue. They eat algae and can live many days without water. A periwinkle has a hinged door in its shell that it can shut, keeping moisture locked inside.

On muddy shores, mangrove forests are often homes for oysters and barnacles, which attach themselves to the trunks and roots of trees. Snails and crabs are common in saltwater marshes.

Food   Invertebrates may eat phytoplankton, zooplankton, or both. Mollusks draw seawater in through their siphons and filter out the tiny creatures, which they then consume. Some invertebrates eat plants or larger animals. Starfish, for example, dine on mollusks.

Reproduction   Most invertebrates are insects, which have a four-part life cycle. The first stage is spent as an egg. The second stage is the larva. It may be divided into several stages between which there is a

Plastic is not a Nutrient

Sea turtles love to snack on jellyfish. Unfortunately, plastic bags puffed with air resemble jellyfish and a sea turtle that can not distinguish between the two, can die. Plastic bags cannot be digested and moved through the animal's body; the bags and other plastic garbage block an animal's digestive tract. Dolphins, seals, whales, and other animals are also in danger from plastics. A gallon-sized plastic milk bottle, a plastic float, a garbage bag, and other items were found wedged in the digestive tract of a dead sperm whale.

The marine iguana lives both on land and in the sea.
IMAGE COPYRIGHT VOVA POMORTZEFF, 2007. USED UNDER LICENSE FROM SHUTTERSTOCK.COM.

shedding of the outer skin casing. A caterpillar is an example of an invertebrate in the larval stage. The third stage is the pupal stage, during which the insect lives in yet another protective casing, like a cocoon. Finally, the adult breaks through the casing and emerges. In most cases, the young are not cared for by the parents. Survival often depends on the absence of predators.

Reptiles   Reptiles, such as lizards and snakes, are cold-blooded vertebrates. Only one species of lizard, the marine iguana, lives both in the sea and on the seashore. Pine lizards may be found in dry areas on sandy beaches where pitch pines grow. Some species of snakes, such as the hognose snake, live among sand dunes on sandy shorelines, preferring the dry conditions found there. Fowler's toads can be found on upper beaches. A saltwater crocodile makes its home in the waters of Southeast Asia, where it lives along muddy seashores near the mouths of rivers. A notable reptile at the seashore is the sea turtle.

Common seashore reptiles   Sea turtles can be distinguished from land turtles by their flippers, which enable them to swim, and by their tolerance for salt water. Green sea turtles are migrators, traveling as far as 1,300 miles (2,094 kilometers) to return to a particular breeding area where they lay their eggs.

The saltwater, or estuarine, crocodile lives in northern Australia and in the region stretching from the east coast of India to the Philippines. It is one of the largest species of crocodiles; average males grow 17 feet (5 meters) long. Most crocodiles float on the surface of the water close to the shoreline where they wait for prey to wander past. The saltwater crocodile is also known to swim out to sea.

Food   Turtles eat soft plants, as well as small invertebrates such as snails and worms. Turtles have no teeth. Instead, they use the sharp, horny edges of their jaws to shred the food enough so they can swallow it.

Sand dune snakes are carnivorous and catch mice and other small prey. They also eat the eggs of nesting shorebirds. Saltwater crocodiles are carnivorous and feed primarily on fish and turtles. They have been known to attack and eat humans.

Reproduction Both snakes and turtles lay eggs. Turtles lay theirs in holes on a sandy shore, which they then cover over with sand. After the nest is finished, the female abandons it, taking no interest in her offspring. Six weeks later the eggs hatch and the young turtles make a run for the ocean and swim away.

Fish   Fish are primarily cold-blooded vertebrates (dependent on the environment for warmth) having gills and fins. Gills are used to draw in water from which oxygen is extracted. Fins are used to help propel the fish through the water.

Most fish found along the seashore are smaller varieties that can live in shallow water or in tide pools. They are dull in color to match the sand or gravel background. Some have suckers (suction cups) on their undersides that allow them to cling to rocks so they will not be washed away. The exception is the flounder, which can grow to be quite large and spends its entire life in the shallow pools of water on the beach.

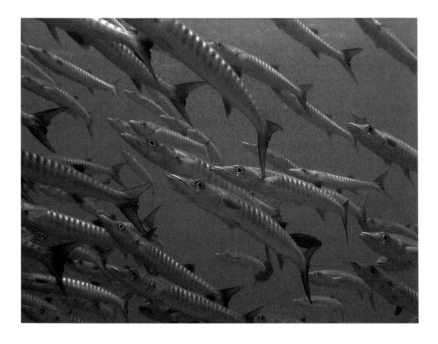

Schools of barracuda cruise the shorelines. They are voracious predators. IMAGE COPYRIGHT SERGEY POPOV V, 2007. USED UNDER LICENSE FROM SHUTTERSTOCK.COM.

Assateague Island to the south. The erosion rate averages 10 feet (3 meters) per year since the walls were built. So far about 9 miles (15 kilometers) of the inlet have been affected.

Overdevelopment　Use of the land for harbors, recreation, and housing has changed the appearance of many shorelines. The popularity of beaches has caused many to become crowded with so many people and buildings that wildlife has either been destroyed or frightened away. In many places, the coastline has lost its natural beauty and has become just another part of a big city.

Dune buggies, dirt bikes, and other recreational vehicles destroy plants and scare animals. Hikers can destroy vegetation, and even well-meaning people who wish only to observe nature can frighten animals or upset the natural rhythms of their lives.

Quality of the environment　In 2006, there were 25,643 occasions when beaches were closed and swim advisories were in effect in the United States because the water was polluted. Most ocean pollution caused by humans is concentrated along the seashore. Sewage and industrial wastes are dumped from coastal cities, adding metals and chemicals to the water. Discarded items, such as plastic bags and food wrappers, pose health hazards for both animals and people.

Insecticides (insect poisons) and herbicides (weed poisons) reach the oceans when the rain washes them from fields and they are carried by rivers to the sea. These poisons often enter the food chain and become concentrated in the bodies of some organisms. Fertilizers and human sewage are also a problem. They cause phytoplankton to reproduce rapidly. When the plants die, their decaying bodies feed bacteria. The bacteria reproduce and use up the oxygen in the water, and other organisms, such as fish, soon die.

Industrial accidents and waste are other problems. It is estimated that 1,435 people died and thousands were paralyzed when they ate fish contaminated with mercury, a metallic element used at a nearby factory at Minamata Bay in Kyushu, Japan in 1953. Oil spills from tanker ships are another danger, as is oil from oil refineries, pipelines, and offshore oil wells. Power plants and some industries often dump warm water into the oceans, causing thermal (heat) pollution. Organisms that require cooler water are killed by the increase in the water temperature.

Agriculture, construction, and the removal of trees digs up the soil. The rain then washes the soil into streams and rivers. Eventually, it enters the ocean to collect as sediment in coastal areas. Some organisms that cannot survive in heavy sediment, such as clams, die. Organisms that inhabited preagricultural, preconstruction, and forested areas also die.

Pollution can travel, so that problems caused by one country may damage the shoreline environment in another.

Native peoples   In most parts of the world, people often settled near coastlines where fish were plentiful and the ocean offered a means of transportation. Since the mid-twentieth century, desirable seashore locations have been taken over by tourists and non-native residents. Few native peoples have continued to live traditional lifestyles. Many have gone to live in cities or adopted more contemporary ways of life. Among those that have tried to maintain important elements of their traditional cultures are the Samoans and Native American tribes of the northwest coast.

World's Worst Oil Spill

Oil spills are especially hazardous to coastlines. The oil is carried ashore by ocean waves and tides, but when the water retreats back to the sea, the oil remains. Thick deposits stick to the shoreline and coat the plants and animals that live there. Oil mats the feathers of birds and the fur of swimming animals leaving the creatures unable to keep warm, and causing many to die of the cold.

The world's worst oil spill took place in 1989 when the ship *Exxon Valdez* was wrecked in Prince William Sound, Alaska. Eleven million gallons (42 million liters) of oil were spilled. Over time, the oil spread out, covering 10,000 square miles (25,900 square kilometers) of ocean and 1,200 miles (1,940 kilometers) of coastline. It is estimated that as many as 6,000 otters, whales, dolphins, and seals, and 645,000 birds died. Three nearby national parks and three national wildlife refuges were also affected.

Kon-Tiki

Norwegian explorer Thor Heyerdahl (1914–2002) was convinced that ancient peoples once sailed across 4,300 miles (6,900 kilometers) of the Pacific Ocean from Peru and colonized Polynesia. To test his theory, he set out with a crew of six people to duplicate the feat in 1947. The expedition lasted 101 days and was carried out on a large raft called the *Kon-Tiki*.

Named for a legendary Inca Sun-god, the *Kon-Tiki* was patterned after sailing rafts used by the ancient Incas, native peoples who lived in what is now Peru. The raft measured 45 feet (13.7 meters) long in the center and tapered to 30 feet (9.1 meters) at the sides. No metal was used.

Instead, the nine thick balsa logs were tied together with hemp rope. Two masts supported a large rectangular sail, and a bamboo cabin was built in the center for shelter.

On April 28, 1947, the raft was set adrift 50 miles (80 kilometers) off the coast of Peru. Ninety-three days later, the *Kon-Tiki* sailed past the island of Puka Puka, which lies east of Tahiti. On August 7, 1947, after the *Kon-Tiki* crashed into a reef in the Tuamoto Archipelago, the voyage was finished, but Heyerdahl had proved his point. The *Kon-Tiki* is now in a museum in Oslo, Norway. Heyerdahl wrote about the voyage in his book, *Kon-Tiki*.

Samoans   The Samoa Islands lie in the Pacific Ocean southwest of Hawaii. The largest island, Savai'i, has an area of only 659 square miles (1,707 square kilometers), so the seashore is easily accessible and an important part of everyone's life.

The people of Samoa are Polynesian and closely related to the native people of Hawaii and New Zealand. Although the islands have felt the influence of Europe and then America since 1722, traditional Samoan culture has remained very important.

The Samoan economy is based on agriculture. Crops include corn, beans, watermelons, bananas, and pineapples. Fish have always been an important food source for local families, but fishing was not commercially important until 1953 when a tuna cannery was built in the town of Pago Pago.

Native Americans of the Northwest Coast   The Columbia and Fraser Rivers in North America open into the Pacific Ocean, and each year millions of Pacific salmon return to those rivers to breed. This ample supply of fish led many native peoples to settle there, including the Tlingit, Haida, Tsimshian, Kwakiutl, Nootka, and Salish. The area proved rich in other wildlife as well. Mussels, clams, oysters, candlefish, herring, halibut, and sea lions were available from the sea. The land provided food, such as

moose, mountain sheep and goats, deer, and many small animals, as well as roots and berries.

Products of the seashore, such as shells, dried fish, fish oil, and the dugout canoes built for transportation were considered a source of wealth. Religious beliefs were based on mythical ancestors whose images were carved on totem poles, boats, masks, and houses.

By 1900, traditional ways of life were disappearing, but the people remained on or near their ancient lands. Many now work in forestry. Some groups are restoring native customs, and there is interest in a return to arts-and-crafts production.

The Food Web

The transfer of energy from organism to organism forms a series called a food chain. All the possible feeding relationships that exist in a biome make up its food web. Along the seashore, as elsewhere, the food web consists of producers, consumers, and decomposers. These three types of organisms transfer energy within the seashore environment.

Phytoplankton are among the primary producers along the lower zone of the seashore. They produce organic (derived from living organisms) materials from inorganic chemicals and outside sources of energy, primarily the Sun. Other primary producers include seagrasses and plants that live in the upper zone.

Zooplankton and other animals are consumers. Animals that eat only plants are primary consumers. Secondary consumers eat the plant-eaters.

Puffins are chunky birds with large bills. They shed the colorful outer parts of their bills after the mating season, leaving a smaller and duller beak.
IMAGE COPYRIGHT TOM CURTIS, 2007. USED UNDER LICENSE FROM SHUTTERSTOCK.COM.

Point Reyes National Seashore
Location: Pacific coast of California, north of San Francisco
Area: 64,546 acres (25,818 hectares)

They include zooplankton that eat other zooplankton. Tertiary consumers are the predators, like starfish and mice that eat the second-order consumers. Some, such as mice and humans, are omnivores, organisms that eat both plants and animals.

Decomposers feed on dead organic matter and include some insects and species of crabs. Bacteria also help in decomposition.

A serious threat to the seashore food web is the concentration of pollutants and dangerous organisms that become trapped in sediments where organisms lower on the food chain feed. These life-forms become food for consumers higher on the food chain, and at each step in the food chain the pollutant becomes more concentrated. Finally, when humans eat contaminated sea animals, they are in danger of serious illness. The same process is true of diseases such as cholera, hepatitis, and typhoid, which can survive and accumulate in certain sea animals, and are then passed on to people who eat those animals.

Spotlight on Seashores

Point Reyes National Seashore   Point Reyes National Seashore is part of a large peninsula (arm of land) extending off California into the Pacific Ocean. Its broad beaches are backed by tall cliffs and forested hills and valleys.

The shoreline along the California coast was carved by volcanoes and earthquakes, and earthquakes are still common. During the 1906 earthquake that struck San Francisco, the Point Reyes peninsula moved more than 16 feet (5 meters) to the northwest. Shifting sands, islands, and steep underwater cliffs are all found in the region.

Wave action is strong and has carved the offshore rocks into rugged shapes. Rip currents and a pounding surf make swimming dangerous in some places. In sheltered areas, quiet bays and lagoons have formed, which are enclosed by sand dunes and grass-covered lowlands.

Sea stars, horseshoe crabs, and other invertebrates live along the shore, and Point Reyes is popular with seabirds, sea otters, and sea lions. Migrating gray whales can often be seen passing in the distance.

Padre Island National Seashore Padre Island National Seashore is located on a barrier island that stretches about 110 miles (177 kilometers) along the coast of Texas in the Gulf of Mexico. Padre Island borders the Laguna Madre, a shallow lagoon. The park consists of sandy beaches and dunes as high as 40 feet (12 meters).

Many waterfowl winter on the island. Other birds include herons, terns, egrets, brown pelicans, and white pelicans. Since the mid-1980s, scientists have been bringing eggs from endangered sea turtles here to help rebuild the population.

The island is uninhabited by people and is the largest undeveloped beach remaining in the United States (exclusive of Alaska and Hawaii).

Virgin Islands National Park Virgin Islands National Park occupies about over 7,000 acres (2,833 hectares) of St. John, the smallest of the three main Virgin Islands, and takes in many tiny islands offshore. Many caves and grottoes are found along the shoreline.

The park's white sandy beaches are backed by steep mountains and valleys covered with tropical forests. The highest point of land is Bordeaux Peak at 1,277 feet (389 meters). Mangrove forests grow along the shore, while a number of coral reefs support tropical fish and other marine creatures such as sea urchins offshore. The only native mammals living within the park are bats.

Pre-Columbian Indians once lived here, and relics of their culture can be found. For two centuries, the islands were a base for pirates, and some people believe that buried treasure may still be hidden in the region.

Acadia National Park Acadia National Park in Maine includes parts of Mt. Desert Island, Isle au Haut, Schoodic Peninsula, and several smaller islands. It is a rugged, rocky shoreline backed by mountains. Coniferous forests grow close to the water. Mt. Desert Island features Cadillac Mountain, which is 1,530 feet (466 meters) high.

More than 10,000 years ago, the shoreline was much further out than it is now. For that reason, the region is often called the drowned coast. Its cliffs were once inland mountains, and the park includes a fjord called Somes Sound. Some areas show the rubble left by glaciers.

Padre Island National Seashore
Location: Coast of Texas, along the Gulf of Mexico
Area: 133,918 acres (53,567 hectares)

Virgin Islands National Park
Location: Caribbean Sea, east of Puerto Rico
Area: 7,000 acres (2,833 hectares)

Acadia National Park
Location: Atlantic coast of Maine
Area: 41,634 acres (16,653 hectares)

Wave action of the North Atlantic is strong here, and huge blocks of granite have been dislodged from the cliffs and lie along the shore. Anemone Cave, a large cavern 89 feet (27 meters) deep, has been carved into the solid rock by wave action. In sheltered areas, small bays and coves have formed.

The intertidal zone has many tide pools and is home to a host of creatures, including sea stars, periwinkles, anemones, sea urchins, and crabs. Seabirds, such as gulls, guillemots, and ducks are common.

Cape Hatteras National Seashore   Cape Hatteras National Seashore in North Carolina is a chain of low, narrow, barrier islands, including Bodie, Hatteras, and Ocracoke. At some points the islands are as much as 30 miles (48 kilometers) off the North Carolina coast. The seashore area has 70 miles (110 kilometers) of beaches, dunes, and salt marshes and is a wildlife refuge.

Cape Hatteras National Seashore is known for severe storms. Off the Cape, which is a promontory (arm of land) on Hatteras Island, cold northern currents meet the warmer Gulf Stream as it moves through the Atlantic. Their collision breeds stormy conditions. Diamond Shoals, a region of shallow water, has been called the "graveyard of the Atlantic" because so many ships have been wrecked there. Their remains are still visible from the shore.

The Cape has had a lighthouse since 1803. The present lighthouse, which is the tallest in the continental United States at 191 feet (58 meters), has been in operation since 1870. Years of severe weather, including hurricanes, threatened to destroy it and it was finally moved inland.

Cape Hatteras National Seashore
Location: Atlantic coast of North Carolina
Area: 30,350 acres (12,282 hectares)

Rhine River Delta
Location: North Atlantic coast of the Netherlands
Area: Approximately 900 square miles (2,340 square kilometers)

Rhine River Delta   The Rhine River Delta in the Netherlands is a vast area connecting the Rhine River with the North Sea. Centuries ago, it was marshy and dotted with small islands. Storms often swept in from the North Sea, and the land was frequently flooded.

Drawn by the plentiful supply of fish, the Dutch began living on the islands in the Rhine River Delta around 450 BC. They built hills of earth on which to live during flood times. In the twelfth century, special walls called dikes were constructed to keep out the sea. Later, Dutch soldiers fighting in Middle Eastern lands learned a trick from the Arabs who used windmills to pump water for growing crops. The Dutch figured the same process could be used in reverse to drain the land. Soon, the Dutch had built dikes around the marshy areas and used windmills to pump the water out into the ocean. By the 1700s, the Netherlands had 10,000 windmills. In later years, steam-driven pumps replaced the windmills, which have become a symbol of the Netherlands and are popular among tourists.

In 1919, work began to reclaim a freshwater lake, called the Zuider Zee, that had been invaded by the sea. A dam was completed in 1932 that allowed the seawater to drain, reclaiming 407,724 acres (165,000 hectares) of land. By 1937, the lake was fresh again and was renamed Ijsselmeer.

A winter storm in 1953 hit and destroyed many dikes along the coast, killing hundreds of people when the land was flooded. The Dutch built more dams across the delta to prevent such an accident from repeating. The project was finished in 1986, closing all the estuaries in the area except two that are used for shipping. Protected by dikes, dams, and pumps, the Rhine Delta's fertile soil is now valuable farmland. About 25 percent of the Netherlands is below sea level during high tide. The lowest spot, 22 feet (6.7 meters) below sea level, lies to the northeast of Rotterdam.

Fjords of Western Norway   Usually found in northern regions along mountainous coasts, fjords are valleys eroded by glaciers. When the glaciers retreated or melted, the ocean poured in, creating long, narrow, deep arms of water that project inland. The fjords of western Norway have high, steep walls; often rising to 3,280 feet (1,000 meters). The crowns of these surrounding cliffs may be heavily forested. Cascading waterfalls sometimes drop from great heights, adding to the spectacular scenery.

The depth of the fjords depends upon how much erosion took place below sea level. The deepest along western Norway is the Sognefjord at 4,291 feet (1,304 meters). Inland, some fjords end in glaciers, from which huge chunks of ice break off creating icebergs.

Fjords are not always penetrated by cold ocean water and are usually ice free in the winter, except along the coast where the water may be more shallow and freezes more rapidly. Gales (high winds) are frequent on the western coast.

Inland fjord valleys are rich in vegetation, especially coniferous trees, lowland birches, and aspens. The coast attracts large numbers of food fishes, such as cod, herring, mackerel, and sprat. Woodcocks and migratory birds, as well as lemmings, hares, red foxes, and reindeer, are common.

Fjords of Western Norway
Location: North Atlantic coast of Norway
Individual Length: As much as 114 miles (182 kilometers)

For More Information

BOOKS

Carson, Rachel L. *The Sea Around Us.* Rev. ed. New York:Chelsea House, 2006.

Cox, Donald D. *A Naturalist's Guide to Seashore Plants: An Ecology for Eastern North America.* Syracuse, NY: Syracuse University Press, 2003.

Harris, Vernon. *Sessile Animals of the Sea Shore.* New York: Chapman and Hall, 2007.

Peschak, Thomas P. *Wild Seas, Secret Shores of Africa.* Cape Town: Struik Publishers, 2008.

Preston-Mafham, Ken, and Rod Preston-Mafham. *Seashore.* New Edition. New York: HarperCollins Publishers, 2004.

Romashko, Sandra D. *Birds of the Water, Sea, and Shore.* Lakeville, MN: Windward Publishing, 2001.

Vogel, Carole Garbuny. *Shifting Shores (The Restless Sea).* London: Franklin Watts, 2003.

Worldwatch Institute, ed. *Vital Signs 2003: The Trends That Are Shaping Our Future.* New York:W.W. Norton, 2003.

ORGANIZATIONS

American Littoral Society, 18 Hartshorne Drive Suite 1, Highlands, NJ 07732, Phone: 732-872-0111, Fax: 732-291-3551, Internet: http://www.littoralsociety.org.

American Oceans Campaign, 2501 M Street, NW Suite 300, Washington D.C. 20037-1311, Phone: 202-833-3900, Fax: 202-833-2070, Internet: http://www.oceana.org.

Center for Marine Conservation, 1725 De Sales St. NW, Suite 600, Washington, DC 20036, Phone: 202-429-5609 Internet: http://www.cmc-ocean.org.

Coast Alliance, P.O. Box 505, Sandy Hook, Highlands, NJ 07732, Phone: 732-291-0055, Internet: http://www.coastalliance.org.

Environmental Defense Fund, 257 Park Ave. South, New York, NY 10010, Phone: 212-505-2100; Fax: 212-505-2375; Internet: http://www.edf.org.

Envirolink, P.O. Box 8102, Pittsburgh, PA 15217; Internet: http://www.envirolink.org.

Environmental Protection Agency, 401 M Street, SW, Washington, DC 20460, Phone: 202-260-2090; Internet: http://www.epa.gov.

Friends of the Earth, 1717 Massachusetts Ave. NW, 300, Washington, DC 20036-2002, Phone: 877-843-8687; Fax: 202-783-0444; Internet: http://www.foe.org.

Greenpeace USA, 702 H Street NW, Washington D.C. 20001, Phone: 202-462-1177; Internet: http://www.greenpeace.org.

Olympic Coast Alliance, P.O. Box 573 Olympia, WA 98501, Phone: 360-705-1549, Internet: http://www.olympiccoast.org.

Sierra Club, 85 2nd Street, 2nd fl., San Francisco, CA 94105, Phone: 415-977-5500; Fax: 415-977-5799, Internet: http://www.sierraclub.org.

World Wildlife Fund, 1250 24th Street NW, Washington, DC 20090-7180, Phone: 202-293-4800; Internet: http://www.wwf.org.

PERIODICALS

Ehrhardt, Cheri M. "An Amphibious Assault." *Endangered Species Bulletin.* 28. 1 January-February 2003: 14.

"Katrina Descends." *Weatherwise.* 58. 6 November-December 2005: 10

Wagner, Cynthia G. "Battles On the Beaches." *The Futurist.* 35. 6 November 2001: 68.

WEB SITES

Discover Magazine: http://www.discovermagazine.com (accessed on September 12, 2007).

Monterey Bay Aquarium: http://www.mbayaq.org (accessed on September 12, 2007).

National Center for Atmospheric Research: http://www.ncar.ucar.edu (accessed on September 12, 2007).

National Geographic Society: http://www.nationalgeographic.com (accessed on September 12, 2007).

National Oceanic and Atmospheric Administration: http://www.noaa.gov (accessed on September 12, 2007).

Scientific American Magazine: http://www.sciam.com (accessed on September 12, 2007).

UN Atlas of the Oceans: http://www.cmc-ocean.org (accessed on September 12, 2007).

Woods Hole Oceanographic Institution? if so: http://www.whoi.edu (accessed on May 14, 2008).

Tundra

In the northern lands close to the Arctic, and on the upper slopes of high mountains all over the world, a unique biome called the tundra is found. The word tundra comes from a Finnish word *tunturia*, which means barren land. In this cold, dry, windy region where trees cannot grow, the often bare and rocky ground supports only hardy, low-growing plants, such as mosses, sedges, heaths, and plantlike lichens (LY-kens), which give it a greenish-brown color. During the brief spring and summer, flowers burst into bloom with the warmth of the sun, dotting the landscape with color.

Tundra covers about 20 percent of Earth's surface. Almost all tundra is located in the Northern Hemisphere (the half of Earth above the equator); small areas do exist in Antarctica in the Southern Hemisphere. Antarctica is much colder than the Arctic and the ground is usually covered by ice. Conditions are seldom right for tundra to form there.

The tundra is not an environment with high biodiversity. (An area with high biodiversity supports a wide variety of plants and animals.) Only a few species of plants and animals live in the tundra, but those few, such as lichens and mosquitoes, are found in great numbers.

The tundra may seem bleak and unfriendly, but it can be a place of eerie beauty, especially in the Arctic during the winter. Winter nights last for weeks and are often lit up by the blue, red, and green colors of the Aurora Borealis, or northern lights. The Aurora Borealis occurs when energy-charged particles from the sun enter Earth's atmosphere and create flashes of light. The Arctic is called the Land of the Midnight Sun because, during the summer, the sun never sets below the horizon and daylight lasts for twenty-four hours.

WORDS TO KNOW

Hypothermia: A lowering of the body temperature that can result in death.

Peat: A type of soil formed from slightly decomposed plants and animals.

Permafrost: Permanently frozen topsoil found in northern regions.

Pingos: Small hills formed when groundwater freezes.

Rhizomes: Plant stems that spread out underground and grow into a new plant that breaks above the surface of the soil or water.

Soredia: Algae cells with a few strands of fungus around them.

Thermokarst: Shallow lakes in the Arctic tundra formed by melting permafrost; also called thaw lakes.

Tussocks: Small clumps of vegetation found in marshy tundra areas.

How Tundra Develops

Tundra forms primarily because of climate. In the Arctic, winters are long and cold, and summers are short and cool. This allows limited plant growth. On high mountains, tundra forms when the location is right to produce the necessary climate.

The lack of soil in a tundra region may be due to erosion (wearing away) from wind and rain. During the Ice Ages more than 10,000 years ago, glaciers scraped away any soil, leaving only bare rock.

Kinds of Tundras

There are two types of tundra: Arctic and alpine. Arctic tundra is found near the Arctic Circle. Alpine tundra forms on mountaintops where the proper conditions exist.

Arctic tundra   Several characteristics are typical of Arctic tundra. One is the polar climate, which has an average July temperature of not more than 50°F (10°C). Arctic tundra is far from the equator. Sunlight hits Earth here at an angle and must pass through more atmosphere. This means the sunlight that reaches the soil contains less energy per square foot (square meter) than at the equator.

Another characteristic of the Arctic tundra is a deep layer of permanently frozen ground called permafrost. Generally, fewer than 18 inches (45 centimeters) of tundra soil thaws during the cool summer. Below that the ground remains frozen. Water from melting snow cannot drain into

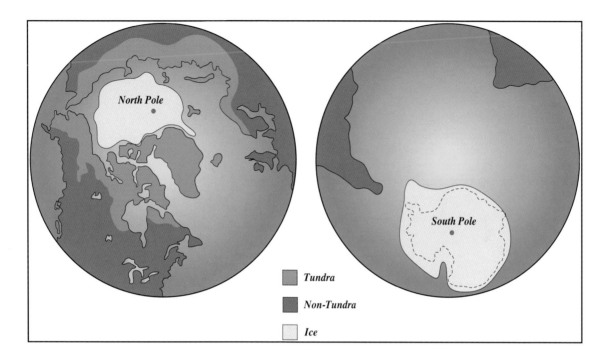

Tundra

Non-Tundra

Ice

the frozen ground, and little evaporates in the cool summer air. As a result, the water becomes trapped on the surface.

Arctic tundra is found on all three northern continents close to or above the Arctic Circle and near the Arctic Ocean. In Asia, Arctic tundra is found in the part of Russia known as Siberia. In Europe, it is found in northern Scandinavia, which includes the countries of Norway, Sweden, and Finland. In North America, Arctic tundra is found in northern Alaska and Canada. Some Arctic tundra is located on islands, such as Greenland and Iceland.

Alpine tundra   Alpine tundra is found at the tops of mountains above the timberline, the point above which trees cannot grow. The timberline and tundra are found at different elevations (heights) in different mountain ranges. The farther the mountains are from the equator, the lower the elevation needed for tundra to form.

Compared to Arctic tundra, alpine tundra gets more rain and its soil drains better because of the sloping terrain. It also gets more sunlight because it is found at lower latitudes (a distance north or south of the equator, measured in degrees) where day and night are more equal in

Storms on the Sun

The Aurora Borealis, or Northern Lights, occur when there are "storms" on the sun. These storms shoot out streams of energy-charged particles called electrons. When the electrons enter Earth's atmosphere and collide with each other, a flash of light occurs. The magnificent Aurora Borealis is created when billions of these collisions occur at the same time.

length than in the Arctic. Usually, there is no lower layer of permafrost in alpine tundra.

Alpine tundra is found in the Rocky, Cascade, and Sierra Mountains in North America, the Andes Mountains in South America, the Alps and Pyrenees in Europe, and the Himalayas in Asia.

Climate

Both Arctic and alpine tundra have very cold climates. Variations in temperature ranges may occur because of location. Arctic tundra located high above sea level (the average height of the sea) has a colder climate. If tundra is found near the coast, ocean currents can affect the temperature. For example, the North Atlantic Drift, a warm ocean current, warms the coast of northern Scandinavia. The coast of northeastern Canada is colder due to the influence of the Labrador Current, an icy current that mixes with the warmer waters. Arctic tundra is windy, with winds ranging between 30 to 60 miles (48 to 97 kilometers) per hour.

Unlike Arctic tundra, alpine tundra has a more moderate climate that varies with latitude and altitude. The farther away from the equator, the colder the temperature becomes. The higher the altitude, the colder and

Elk can be seen on the alpine tundra. IMAGE COPYRIGHT GEORGE BURBA, 2007. USED UNDER LICENSE FROM SHUTTERSTOCK.COM.

windier the climate. At 15,000 feet (4,500 meters) above sea level, the climate changes as much as if it were located 10° latitude farther north. High altitude also means that the atmosphere is thin, which results in little oxygen present in the air.

Temperature   The Arctic tundra is one of the driest and coldest biomes on Earth. It is generally described as a cold desert. January temperatures average from –4° to –22°F (–20° to –30°C). In Siberia, an extremely cold area in north central Asia, winter temperatures may drop to –40°F (–30°C) or lower. Scandinavian tundra is relatively warm, and the winter temperatures may average 18°F (–8°C). Average arctic temperatures in summer range from 38° to 50°F (3.3° to 10°C). Although daytime temperatures above 90°F (32°C) have been recorded, normally the average does not rise much above 50°F (10°C) because temperatures drop significantly in the evenings.

Antarctica's small tundra region is even colder, with an average annual inland temperature of –70°F (–56°C).

Alpine tundra is cold, but temperatures are more moderate than in the Arctic. Elevation affects the temperature in an alpine tundra, and mountains have been described as "open windows letting the heat out." Scientists estimate that for every 1,000 feet (305 meters) in height, the temperature drops 3.6°F (–15.8°C).

Winter temperatures in an alpine tundra rarely fall below 0°F (–18°C) and summers are cool. The average annual temperature in the Peruvian mountains seldom falls below 50°F (10°C). On Alaskan mountains, temperatures in January average about 8°F (–13°C) and almost 47°F (8°C) in July.

Precipitation   Arctic tundra is similar to a cold desert in that it receives little precipitation (rain, snow, or sleet), usually fewer than 10 inches (25 centimeters) annually. Most precipitation falls as snow in winter.

Alpine tundra receives more rain than Arctic tundra, but the water runs rapidly off the mountain slope, leaving any dry soil to blow away in the wind.

Arctic Heat Wave

Sometimes the weather can become quite warm during summer on the Canadian tundra, warm enough for people to swim in the Arctic Ocean. In 1989 the temperature rose above 90°F (32°C) at Coppermine in the Northwest Territories. These temperature increases are caused by warmer water brought into the Arctic Ocean by northward-moving currents, such as the Irminger current. This current is a branch of the Gulf Stream, which is a warm ocean current that begins in the Gulf of Mexico.

Geography of Tundras

The excessively cold temperatures affect the geography of the tundra.

Landforms   In Arctic tundra, water trapped in the top layer of soil does strange things to the landscape. The water freezes each winter, and the ice melts each summer. When water freezes, it expands (takes up more space). When ice melts, it contracts (shrinks). This yearly expansion and contraction cracks and breaks rocks, creating hills, valleys, and other physical features.

A pingo is a small circular or oval hill formed when a pool of water under the ground freezes and forces the soil up and out. The hill may grow a few inches taller every year. Some pingos are as high as 300 feet (90 meters) and more than half a mile (800 meters) wide. Stone circles are formed by piles of rocks that have been moved into a more or less circular shape by the expansion of freezing water.

Polygons are cracks in the ground that take on geometric shapes because of the freezing and thawing action. If the soil is rocky, the rocks can be pushed up through the cracks, making the geometric shapes even more distinct.

Irregularly shaped ridges, called hummocks or hammocks, are formed when large blocks of ice meet and one slides over the top of the other. When the ice melts, the ground is uneven. Their heights range from 65 to 100 feet (20 to 30 meters).

Areas of bare, rock-covered ground on alpine tundra are called fell-fields. They are often formed when rock and soil slide down a slope.

Elevation   Most of the low, rolling plains of the Arctic tundra are located about 1,000 feet (300 meters) above sea level, whereas alpine tundra is found high on mountains above the tree line. In northern latitudes, close to the Arctic Circle, arctic tundra may begin at about 4,000 feet (1,200 meters) above sea level. In more southern mountains, alpine tundra usually begins around 12,000 feet (3,657 meters) above sea level.

Soil   The soil in Arctic tundra has two layers. The surface layer is called the active layer because it freezes in winter and thaws when the weather warms up. This layer is shallow, its depth ranging from about 10 inches to 3 feet (25 centimeters to 1 meter). About 15 percent of the active layer is well drained because it is located on stony or gravelly material, on slopes,

Permafrost lake in tundra above the Arctic Circle in Canada. IMAGE COPYRIGHT OKSANAPERKINS, 2007. USED UNDER LICENSE FROM SHUTTERSTOCK.COM.

and in elevated areas. The remaining 85 percent of the active layer is usually poorly drained and remains wet.

The lower, or inactive, layer of soil stays frozen throughout the year and is called permafrost. Permafrost prevents the water captured in the active layer of soil from draining away. Made of such materials as gravel, bedrock, clay, or silt, permafrost reaches depths of 300 to almost 2,000 feet (90 to 609 meters). In Russia on the Taimyr Peninsula, permafrost goes very deep, reaching 1,968 feet (600 meters), whereas permafrost on tundra near Barrow, Alaska, descends to only 984 feet (300 meters).

Tundra soil is generally poor in nutrients, such as nitrogen and phosphorous. In some areas where animal droppings are plentiful and fertilize the soil, vegetation is lush. Near the southern edge of the Arctic tundra, for example, the soil can be boggy. Bog soil contains little oxygen, is acidic, and is low in nutrients and minerals.

Treasures in the Ice

The remains of plants and animals millions of years old have been discovered in tundra permafrost. Tree stumps found on Canadian tundra could still be burned for fuel.

The bodies of many mammoths have been perfectly preserved in the Siberian tundra. These large animals, now extinct, are related to modern-day elephants. In the early 1900s, a mammoth was discovered with its head sticking out of a bank of the Berezovka River. Although wolves had eaten part of his head, its tongue and part of its mouth were still preserved. In its mouth, between its teeth, were the remains of the sedge and buttercups it had been eating. Both of these plants still grow on the tundra.

Alpine tundra rarely has a permafrost layer, but the soil does freeze and thaw as in the Arctic. When permafrost does occur in alpine tundra, it is at higher elevations where temperatures are colder and in areas where mud, rock, and snow slides are common.

Similar to arctic soil, alpine soil is stratified (arranged in layers). Alpine soil has good drainage because of the sloping terrain. However, fierce winds may dry it out and blow it away. Some alpine regions are covered with material so weathered and thin it cannot be classified as soil.

Water sources Melting snow in Arctic tundra has nowhere to go since it cannot sink into the ground, and temperatures never get warm enough for it to evaporate. As a result, in summer the tundra is covered with marshes, lakes, bogs, and streams. Marshes are a type of wetland characterized by poorly drained soil and plant life dominated

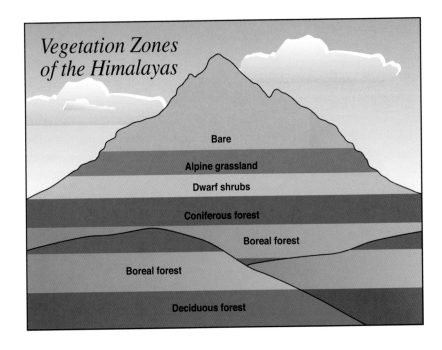

Vegetation Zones of the Himalayas

Bare

Alpine grassland

Dwarf shrubs

Coniferous forest

Boreal forest

Boreal forest

Deciduous forest

by nonwoody plants. Bogs are a type of wetland that has wet, spongy, acidic soil, called peat.

Thermokarst, or thaw lakes, are shallow bodies of water unique to Arctic tundra and formed by melting ground ice. Permanent rivers flow into tundra, like the Mackenzie and Yukon rivers in Alaska, and the Lena, Ob, and Yenisei in Siberia. These rivers are partially or completely covered by ice for about six months of the year.

In alpine tundra, water from melting snow and glaciers usually runs off the slopes. In areas where depressions in the ground occur, ponds and marshes form. Mountain streams, which flow during the five warm months of the year, are formed by surface runoff and from springs.

> ## Nature's Hothouses
>
> Since the tundra growing season is so short, some plants get a head start by making use of hothouses formed by the sun. Darker colors absorb more heat, so the sun melts some snow close to the dark soil. This forms small caves in the snow. The floor is the soil and the roof is a dome of snow that remains frozen. Poppies and saxifrages grow well in these miniature hothouses because the air inside is warmer than the outside air.

Plant Life

Permafrost and the yearly freezing and thawing, break up plant roots and make it impossible for trees and other tall plants to survive on Arctic tundra. Plant growth in the alpine tundra is also affected by freezing and thawing. The sloping terrain, exposure to more light, and the lesser amount of moisture available affect the kinds of plants that grow in mountainous alpine tundra.

Despite the harsh conditions, about 1,700 kinds of plants grow on tundra, including sedges, reindeer mosses, liverworts, and grasses. The 400 varieties of flowers found add a wide range of colors.

Algae, fungi, and lichens   It is generally recognized that algae (AL-jee), fungi (FUHN-ji), and lichens do not fit neatly into the plant category.

Algae   Algae play an important role in the tundra. Most species have the ability to make their own food by means of photosynthesis (foh-toh-SIHN-thuh-sihs); the process by which plants use the energy from sunlight to change water and carbon dioxide from the air into sugars and starches. Other species of algae absorb nutrients from their surroundings. In the Arctic, algae are found in the ocean and wetland areas. In alpine tundra, algae grow on unmelted snow. These algae have a red pigment.

Algae may reproduce in one of three ways. Some split into two or more parts, with each part becoming a new, separate plant. Others form spores (single cells that have the ability to grow into a new organism). A few reproduce sexually, during which male and female cells unite to create a new plant.

Fungi   Fungi are plantlike organisms that cannot make their own food by means of photosynthesis; instead they grow on decaying organic (derived from living organisms) matter or live as parasites on a host. Fungi are decomposer organisms that, together with bacteria, are responsible for the decay and decomposition of organic matter. One of the most important roles of fungi in tundra is formation of lichens.

Fungi form spores to reproduce. These spores are carried from one location to another on the air or by animals.

Lichens   Lichens are the most common tundra plant, with about 2,500 species growing in Arctic and alpine regions. Lichens are combinations of algae and fungi living in cooperation. The fungi surround the algae cells. The algae obtain food for themselves and the fungi by means of photosynthesis. It is not known if the fungi aid the algae organisms, but they may provide them with protection and moisture. In harsher climates, lichens are often the only vegetation to survive. They have no root system so they can grow on bare rock, adding beauty to the tundra with their colors of orange, red, green, white, black, and gray.

Despite the short growing season, lichens thrive in their harsh environment. They freeze in winter, but continue to grow in spring. Lichens often live for hundreds of years, although growth is slow.

Like algae, lichens can reproduce in several ways. If a spore from a fungus lands near an alga these two different organisms can join together to form a new lichen. Lichens can also reproduce by means of soredia (algal cells surrounded by a few strands of fungus). When soredia break off and are carried away by wind or water they form new lichens wherever they land.

Reindeer lichen, also called reindeer moss, is one of the most common tundra plants. It provides a key source of food for Arctic plant-eating animals like caribou and reindeer. Another common type of lichen, called the British soldier, has a tall stalk with a red cap on top, which makes it resemble its namesake.

Green plants   Most green plants need several basic things to grow: light, air, water, warmth, and nutrients. Light, air, and water provide almost all of their needs. The remaining nutrients, primarily nitrogen, phosphorus, and potassium, are obtained from the soil. While water is plentiful for tundra plants, warmth, which brings on the growing season, is available for only a short period of time.

Plants in the tundra grow low to the ground. This helps them stay warm and protected from the high winds. Since the tundra receives little precipitation, many plants have nearly invisible hairs on their leaves, stems, and flowers that reduce moisture loss. Their roots form a dense mat just under the surface of the ground enabling them to quickly draw moisture from melting snow and store it in their leaves. Many plants grow in a tight clump, which helps traps heat. Tussocks are thick clumps of plants, such as cotton grass, that grow to about 1 foot (30 centimeters) in height and are found in marshy areas.

Cozy Flowers

Some tundra insects find warmth and shelter by hiding inside flowers. Inside a buttercup, for example, the temperature can be 40°F (4°C) warmer than the outside air. During the day, these flowers lean toward the sun, following its path to get as much heat as possible. For that reason, they are called heliotropes, which means "to turn toward the sun."

Some alpine plants have deeper roots that help prevent soil erosion. Many alpine plants have red leaves because of a coloring matter called anthocyanin. This special pigment protects plants from the dangers of ultraviolet radiation (a form of harmful light within the sun's rays). Twice as much ultraviolet radiation reaches alpine tundra than regions at sea level because of the high elevation.

Common green plants   Typical flowering plants on Arctic tundra include Arctic lupine, yellow poppy, saxifrage, Arctic campion, Lapland rhododendron, buttercup, campanula, and barberry. The Labrador tea, a hardy evergreen, is very common. Non-grass perennials, called forbs, lie dormant all winter with their growth buds protected underground.

The rosette plant is well-designed to survive in harsh Arctic weather. This type of plant forms rings of leaves around a central growth bud. The leaves protect the fragile bud and help to trap insulating snow in winter and dew during the dry growing season.

Examples of alpine tundra plants include sorrel, saxifrage buttercup, plane leaf, tufted hair grass, alpine bluegrass, and alpine sandwort. The alpine azalea is a member of the heath family called a cushion plant. Cushion plants grow in groups, tightly clumped together, so the plants on the outer edge can protect the ones in the middle. At the lower edges of alpine tundra grow Krummholz, dwarf trees kept small by the cold, icy, windy environment.

The small area of tundra located on the Antarctic Peninsula is home to the only two flowering plants found on that continent, pearlwort and hair grass.

Labrador tea, a hardy evergreen, is common on the tundra. IMAGE COPYRIGHT MASLOV DMITRY, 2007. USED UNDER LICENSE FROM SHUTTERSTOCK.COM.

Growing season   The growing season on tundra is short, lasting from six to ten weeks in June, July, and August. In the Arctic, during those weeks the sun shines almost all day. As a result, plants must make the most of the opportunity for growth.

Green plants may be annuals or perennials. Annuals live only one year or one growing season. Perennials live at least two years or two growing seasons. The above ground portion of perennials often appears to die in winter, while the roots remain dormant underground. The plant returns to "life" in the spring when the weather becomes warmer. Most tundra plants are perennials that can flower quickly and take advantage of the short growing season.

Reproduction   Perennials reproduce and spread by forming seeds. Some plants are self-pollinating, which means that the male and female reproductive cells come from the same plant. Most perennials need pollen from another plant. This process is aided by insects attracted to the colors of the flowers and that travel between flowering plants transferring the pollen from the male reproductive part of a plant, called the stamen, to the female reproductive part of another plant, called the pistil. Strong winds help scatter the fertilized seeds.

Some plants in the more northern, colder Arctic tundra reproduce by budding and division rather than by flowering since the growing season is short. In budding, a new plant develops from any part of the parent. In division, a piece of the parent plant breaks off and develops into a new plant. Forbs reproduce in this way. Mountain sorrel reproduces through rhizomes (RY-zohmz), rootlike stems that spread out under the soil and form new plants.

Endangered species   Tundra plants are very fragile. In the Arctic, traffic during construction of the Alaskan oil pipeline that began in the 1970s damaged the tundra, which did not recover until long after construction had ended. In alpine tundra, beautiful plants are endangered because people pick them or step on them. Their short growing season makes it difficult for them to be easily replaced.

The Penland alpine fen mustard, a small perennial with white flowers found on the Rocky Mountain tundra in Colorado, grows on wetlands fed by melting snowfields. This little plant is endangered because mining and off-road vehicles, which leave tracks in the soil, divert the flow of water away from its habitat.

Heather and lichen vegetation. IMAGE COPYRIGHT VERA BOGAERTS, 2007. USED UNDER LICENSE FROM SHUTTERSTOCK.COM.

Animal Life

The tundra is a permanent home to only a few species of animals because of its harsh environment. Birds, caribou, and red deer, for example, spend only the summers there. The Antarctic tundra has the fewest animals.

For tundra animals, size is an important factor in preventing heat loss. When an animal's appendages (arms, legs, tails, ears) are small, they lose less heat. The Arctic fox, for example, has small ears, short legs, and a short tail. This means there is less area from which body heat can escape.

Microorganisms   Microorganisms are small organisms, such as a protozoan or bacterium, that cannot be seen with the human eye. Bacteria live in the active layer of tundra soil and help decompose dead plants and animals.

Invertebrates   Invertebrates are animals that do not have a backbone. Clams, mussels, snails, crabs, and shrimp are invertebrates found in marshy areas. The primary invertebrate population of tundra consists of a few species of insects that are present in large numbers. There are more mosquitoes on the Arctic tundra, for example, than anywhere else on Earth, because the wet summers provide perfect breeding conditions. As many as one million mosquitoes can be found in an area 1 square yard (0.8 square meter). Springtails are the dominant organism of tundra soil. Populations can reach in excess of several billion per acre (0.405 hectare).

Arctic insects are darker in color than insects elsewhere, which is better for absorbing heat from the sun. They also have more hair to help them conserve heat and energy. Many tundra insects do not have wings.

Arctic tundra moss and berries turn red and green in autumn.

Some scientists believe this helps them conserve energy. Others believe it is because they would not be able to fly in the strong winds. The springtail, for example, has a springlike appendage on its stomach. It hits the ground with this spring and then bounces to a new location, much like a person can on a pogo stick. Those that can fly, stay close to the ground so they do not get blown away.

Common invertebrates Two common tundra insects are the aphid and the midge. Aphids, also called plant lice, feed on the sap from willows, saxifrages, sedges, and other plants. Most aphids are wingless females. During the warmer weather, the females produce live young, which are also all females, without mating. In the fall, they give birth to a generation that includes both males and females. After mating, the females of this generation lay eggs that hatch in the spring to start new colonies.

Midges are small insects with two sets of wings. One set is for flying and the other for balance. Unlike the mosquito, midges do not bite. They are so numerous that their mating swarms are said to look like tornado clouds and can obstruct the vision of any animal they encounter. Midge larvae live in watery areas and feed on algae and dead matter. Midges can grow very old, some living as long as six years.

While Arctic tundra is infamous for its mosquito population, alpine tundra is known for butterflies and biting flies. In the small area of tundra in Antarctica, spiderlike mites and wingless springtails are found. The largest animal living here is the midge, which is half an inch (1.27 centimeters) long.

Food Some insects eat plants and some eat other insects or animals. Male mosquitoes, for example, feed on plant juices, while females feed on the blood of animals and humans. During the summer, plenty of water is available in Arctic tundra because the surface water cannot sink into the frozen ground. In alpine tundra, surface water and streams are sources of water.

Reproduction   The first part of an insect's life cycle is spent as an egg. The second stage is the larva, the immature or young stage of a developing insect. This second part of the life cycle may be divided into several steps between which the developing insect grows larger and sheds an outer skin casing. Some larvae store fat in their bodies and do not need to seek food. During the third, or pupal, stage, the insect's casing offers as much protection as an egg. Finally, in the last stage, the adult emerges. These stages can occur quickly or take up to several years. In the colder tundra regions the process is slow. The midge, for example, takes two years to complete its life cycle in tundra, but only six months in warmer climates.

In order to mate, some wingless, female insects in tundra emit a scent into the air that attracts males, who can fly. In species where both males and females fly, the insects swarm in great numbers. Some males are able to recognize females by the vibrations given off by the beating of their wings. In some species of Arctic insects, like the caddis fly and midge, males are few. Females carry all the necessary genetic information and reproduce without having to mate with a male. The eggs are stimulated to develop by chemical means rather than by male sperm.

Amphibians   Amphibians, such as salamanders and frogs are vertebrates (animals with a backbone) that live at least part of their lives in water. Most amphibians are found in warm, moist, freshwater environments and in temperate zones (areas in which temperatures are seldom extreme). Only a few amphibians live in Arctic tundra and none are usually found in alpine tundra.

Common amphibians   The Siberian salamander is the only salamander found in the tundra. This hardy creature spends winter frozen in the permafrost in Russia and has been found alive at depths of almost 50 feet (14 meters).

The Hudson Bay toad lives in Arctic tundra in North America. The wood frog, which is common in Alaska, is sometimes found on the tundra as well.

Food   Amphibians use their long tongues to capture prey. Even though they have teeth they do not chew, but swallow their food whole.

Mosquito Antifreeze

Mosquitoes can keep themselves from freezing in winter the same way people keep their cars from freezing—with antifreeze. Some mosquitoes replace the water in their bodies with a chemical called glycerol. Their bodies are able to manufacture glycerol from other substances such as fat. With this protection, they can spend the winter under the snow and live to bite again the following summer.

Graylings, another relative of the salmon, live in tundra streams. Pike and trout spend some time in tundra.

Reproduction   Fish reproduce by laying eggs. Usually, the female releases eggs and the male releases sperm and fertilization takes place in the water. The eggs can cling to rocks or float on the water's surface. Some species dig depressions in the ground under the water and deposit eggs there.

Salmon spend three to five years at sea before they return to the same stream where they were born to spawn (reproduce). When spawning is complete, they die.

Birds   Birds are vertebrates. Hundreds of species of birds visit Arctic and alpine tundra, but few stay all year long. The snowy owl, the raven, the willow ptarmigan (TAHR-mah-guhn), and the rock ptarmigan are permanent residents of the Arctic. One of the only species of birds found in both the Arctic and alpine areas is the water pipit.

Many species of birds migrate to the Arctic tundra to breed. They fly south in winter and return to the tundra in summer with their mates to build nests and lay their eggs.

Common birds   Water birds that migrate to the Arctic tundra include geese, ducks, and even a few swans. The Arctic tern makes the

The water pipit is the only species of bird found in both the Arctic and alpine tundras.
IMAGE COPYRIGHT CARLOS ARRANZ, 2007. USED UNDER LICENSE FROM SHUTTERSTOCK.COM.

The Ptarmigan is one of the few permanent residents of the Arctic. IMAGE COPYRIGHT TTPHOTO, 2007. USED UNDER LICENSE FROM SHUTTERSTOCK.COM.

longest journey, flying from Antarctica in a round trip of at least 25,000 miles (40,000 kilometers).

Birds that summer on Arctic tundra include gulls, plovers, redpolls, buntings, loons, warblers, red phalaropes, and skuas. Some of these birds eat insects, which are plentiful during the summer. The skua and the gyrfalcon are predators that eat other birds and small animals.

Alpine tundra is home to many birds. Some species of hawks, eagles, and falcons live above the tree line and eat small mammals, such as rodents and hares, or smaller birds. Wall-creepers eat insects and may move lower down the mountain in winter. Alpine cloughs, wall-creepers, and accentors stay close to the ground to avoid the winds. Some finches remain at higher elevations all year.

The ptarmigan is a brownish-colored grouse with feathered legs and feet. It is one of the few birds that spends nearly all year on the tundra, both in Arctic and alpine regions. Its feathers grow very dense and turn white during the winter. Sometimes ptarmigans sleep in the snow in order to preserve body heat, which is more easily lost out in the open. The ptarmigan is a plant-eater and, in very severe winter weather, travels below the timberline in search of food.

Food   The marshy areas of tundra provide plenty of water for all animals, and many birds feed on water insects. Plant material, shellfish, and carrion are other sources of food. The golden eagle feeds on small mammals, usually rabbits. It will also eat other birds and carrion.

The Monster of the Mountains

The Sherpa people of Nepal believe they share their mountain home with a creature even larger than the yak—the Yeti, or abominable snowman. A yak can weigh up to 2,200 pounds (1,000 kilograms), which would make a Yeti huge. Many footprints have been found and a number of Sherpa claim to have seen one, yet no Yeti, dead or alive, has ever been proven to exist.

Reproduction   All birds reproduce by laying eggs. Usually, male birds must attract the attention of females by singing, displaying feathers, or stomping their feet. After they mate and the female lays her eggs, she sits on them to keep them warm until they hatch. Two species of Arctic swimming birds, the red phalaropes and the northern phalaropes, reverse these roles. The female shows off to attract the attention of the male. After they mate, she usually lays four eggs. The male sits on them until they hatch and then he cares for the young.

Some birds, like the least sandpiper, take advantage of the short breeding season by rotating parental duties. After the female lays the eggs, the male mates with one other female and then returns to the nest of the first female to take care of the young. This frees the female to mate again with another male, ensuring that enough young birds will be born to survive the cold weather.

Many birds, like the ruddy turnstone, shelter along rocky coasts and shores. Their nests are hidden in depressions in the ground, called scrapes, that they line with tundra grasses.

Mammals   Mammals are warm-blooded vertebrates (having a backbone). They maintain a consistent body temperature, are covered with at least some hair, and bear live young that are nursed with milk.

In a barren land such as tundra, animals that live there all year long must be very clever in seeking shelter from both predators and the cold. Mammals adapt in several ways. Many, like the Arctic fox, grow a thick, insulating cover of fur, which not only keeps them warm, but also helps to camouflage them. Their coats turn white in winter and brown in summer so that they blend in with the environment making it hard for predators to see them. Most mammals accumulate deposits of fat under their skins, which insulates them and provides a source of nourishment when food is scarce. Some mammals, like the Alaskan marmot and the Arctic ground squirrel, wait out the cold by hibernating until the weather turns warm.

Alpine mammals must adapt to mountain living as well as the cold. Ibexes, for example, have soft pads on their feet that act like suction cups to help them grip the steep rocks. Since there is less oxygen high in the mountains, the yak and mountain goat have developed large hearts and lungs and have more red blood cells (which absorb oxygen) than would be found in similar animals living closer to sea level. These features allow their bodies to use oxygen more effectively. Alpine mammals often seek shelter in forests below the tundra or in rock caves underground. Yaks and musk oxen can withstand the worst cold the tundra has to offer.

Common mammals About forty-eight species of land mammals live on tundra. Large Arctic mammals include musk oxen, caribou, barren-ground grizzly bears, wolves, and polar bears. Smaller Arctic mammals include hares, lemmings, and squirrels. Large alpine mammals include mountain goats, wild sheep, ibexes, red deer, and snow leopards. Pikas, marmots, chinchillas, hares, and viscachas are among the smaller alpine mammals.

The Arctic fox is found in both alpine and arctic tundras.
IMAGE COPYRIGHT SAM CHADWICK, 2007. USED UNDER LICENSE FROM SHUTTERSTOCK.COM.

Food Mammals are either meat-eaters or plant-eaters. Musk oxen and caribou are among the tundra's plant-eaters. Reindeer lichen, is the bulk of the caribou's diet. Other vegetarians include lemmings, hares, and squirrels.

The barren-ground grizzly and the polar bear are primarily meat-eaters, feeding on seals, birds, and fish. The polar bear supplements its diet with seaweed and grass. Wolves eat hares, musk oxen, and caribou. Foxes eat lemmings, stoats, ptarmigans, and hares.

Reproduction Mammals give birth to live young that have developed inside the mother's body. All young are nursed with milk produced by the mother and must remain close to her until they are able to find their own food. Some mammals, like lemmings, are helpless at birth, while others, like caribou, are able to walk and even run almost immediately. Some are even born with fur, and with their eyes open.

Polar bears give birth in dens under the snow in the Arctic tundra. Caribou calves are born in the open as the animals migrate north in spring to their tundra feeding grounds. Musk oxen are also born on open tundra.

Caribou Caribou are migratory herd animals that spend their summers in the Arctic and their winters in the forest on the edge of tundra. Their winter coat is long and thick to protect them from freezing. Their hooves are wide and flat for easy movement over the snow and ice.

Caribou is a Native American word that means "wandering one." Caribou migrate farther than any other land animal, some traveling over

Lemmings on the March

Lemmings are small plant-eating mammals related to field mice. During winter they live in the shelter of burrows in the snow, feeding on seeds and plants. In years when there is a good food supply, lemmings reproduce rapidly. One female may give birth four or five times a year and have six to ten offspring at a time. After three to five good years the tundra is swarming with lemmings. Before long, there is not enough food for all of them. Suddenly, small groups of lemmings begin to run north.

As the lemmings run, they are joined by more and more of their fellow creatures until thousands are on the march. They cross mountains and rivers, never stopping to eat. Many starve to death along the way and many more are eaten by predators.

When the survivors reach the Arctic Ocean they jump in the water. Some scientists say it is an attempt to cross it, some say it is to get food. The tundra seems empty, but the lemmings that stayed behind soon repopulate it.

1,200 miles (2,000 kilometers) each year. They are found in Alaska, the Yukon, and the Northwest Territories of Canada.

In summer, caribou feed on birch leaves and grasses. In the winter they feed on lichens, a staple in their diet, and any remaining twigs or tree buds.

Caribou give birth to their young on the Arctic tundra, usually in the same area the herd has used for many years. After only three days, a calf is strong enough to travel with the adults.

Pika The pika is a small, short-legged animal that resembles a cross between a guinea pig and a rabbit. An adult pika weighs about 5 ounces (140 grams), has rounded ears, a stocky body, and almost no tail. Pika's have sharp, curved claws to help it climb on rocks, and the pads of their feet have thick fur to ensure stable footing when they leap from rock to rock. Protected against the cold by its thick fur, a pika can remain active all winter.

Pikas have two sets of upper teeth, one behind the other. These teeth are sharp and used for tearing the plants they eat for food. They produce two different types of droppings. One is a special, jellylike pellet the animal eats to gain extra nutrients that may not have been absorbed the first time the food went through its digestive tract. The second is a normal solid pellet of unneeded waste matter.

Pikas are found in rocky areas, usually at high altitudes. Colonies are formed around large boulders or rock slides where vegetation can be found nearby. They collect plant materials in the summer and make "haystacks," which are their winter food stores. These haystacks consist of their chief food-plant, the avens, a member of the rose family.

Male and female pikas mark an area in which they live all year. They have two to six young in a litter and produce two litters every summer. Females generally do not leave their territories, but the males roam in search of food.

Pikas are most common in Asia, especially northeast Siberia, but a few species are found in Arctic tundra and alpine tundra in Alaska.

Musk oxen When winter comes to the Arctic tundra, most animals head for shelter. Some go south while others go underground. Musk oxen do neither. Their thick coats keep them warm in the worst Arctic weather, and they remain on the open tundra all winter long, foraging under the snow for plants to eat. In spring, musk oxen shed their warm undercoat, called qiviut. This hair is highly prized by local women who gather it as it is shed each year to knit it into garments, such as scarves, to be worn or sold.

Wolves Wolves live together in familylike packs of five to twenty members on both Arctic and alpine tundra. They are sociable animals, which helps the pack remain together. Each pack is led by a male and female called the alpha (lead) pair. These two wolves produce the yearly litter and other members of the pack help care for the pups. Pack members may watch the young so the mother can join in a hunt. When a kill is made, a wolf can store partially digested meat in its stomach. This is then regurgitated (vomited) and fed to the pups or is shared with a nursing mother.

Endangered species Overhunting and overfishing are two of the most common reasons animals in tundra are endangered. Their habitats are also being disrupted as more people move into the area. Some animals, like the snow leopard and the chinchilla, are rarely seen in the wild, but are raised in captivity in zoos or on fur farms.

The Eskimo curlew is a dark-colored shorebird with a long, downward-curving bill. At one time, the nesting grounds of the curlew were on Arctic tundra of Alaska and Canada and on the coast of the Chukotka Peninsula in Siberia. The bird is endangered because of unrestricted hunting. Its habitats have been destroyed as land is cultivated for farming and used for grazing herds. Scientists do not know

The Warmest Coat on the Tundra

The chinchilla is a small South American rodent that lives high in the Andes Mountains. It has one of the warmest coats found in nature, a beautiful blue-gray fur. Usually, a mammal grows only one hair per pore, but chinchillas have as many as sixty hairs growing in each pore. As a result, their fur is extremely thick and soft, keeping the chinchilla warm in the cold, even during the freezing alpine nights.

Pikas are mammals found in the rocky tundra regions of Asia, especially Siberia. IMAGE COPYRIGHT SERG ZASTAVKIN, 2007. USED UNDER LICENSE FROM SHUTTERSTOCK.COM.

The polar bear is rapidly becoming a symbol of global warming because of the thinning of sea ice, which endangers their survival.
IMAGE COPYRIGHT FLORIDASTOCK, 2007. USED UNDER LICENSE FROM SHUTTERSTOCK.COM.

the location of its current breeding grounds. The Eskimo curlew is close to extinction, if not extinct already.

Caribou in Greenland and the Peary Caribou in the Canadian Arctic are two groups whose numbers are low. Caribou are affected by predators such as wolves and hunters, as well as changes in their habitats, including the building of natural gas pipelines that cut across their migration routes.

Human Life

Few people live on Arctic tundra; those who do are spread all around the polar region. Most people who live in areas of alpine tundra make their homes below the timberline. In some areas, people live below the timberline in winter and move onto the tundra in summer. The Kohistani people of Pakistan, for example, move from 2,000 feet (600 meters) in winter to 14,000 feet (4,300 meters) in summer. Swiss farmers move their goats to the tundra to graze during summer. While there, the farmers raise some crops and make butter and cheese.

Although some native peoples continue to live a traditional way of life, most tundra peoples have adopted more modern lifestyles.

Impact of the tundra on human life   Arctic tundra is harsh land, and life is shaped by the bitter cold. For humans, a few minutes' exposure to the cold can lead to frostbite and the possible loss of fingers, toes, noses, or ears. Longer exposure can lead to hypothermia (a lowering of the body temperature) and death.

There is very little people can do to physically adapt to tundra conditions, instead, they must change their behavior. This is done by living a lifestyle in which nature is respected, people are creative, and nothing is wasted. For example, when the Inuit peoples of North America kill a caribou for food, they use the skin for tents, bedding, and clothes. Harpoons are fashioned from the antlers, and needles and other tools are made from the bones. Thread for sewing is made from the animal's tendons.

Food   Unlike people in more temperate and fertile biomes where the growing season is much longer, the people of the Arctic tundra do not rely on growing grains for food. They find their food by hunting and gathering. Some peoples, such as the Sami, who live in Scandinavia and

Russia, survive by herding animals. Their diet consists primarily of meat from caribou and freshwater fish like char. The ocean is another source of food, and they hunt whales, walruses, and seals. Berries and herbs supplement the diet, and leaves are used for medicine. The traditional diet may be supplemented by canned goods shipped in during the summer months.

Alpine tundra dwellers hunt wild mountain animals for food, such as sheep, goats, and red deer. They also use domesticated llamas and yaks as food sources. South American Indians raise llamas for milk. In the milder alpine climate crops such as barley and potatoes are grown.

Shelter Traditionally, people of the tundra have built their homes from the materials at hand, mainly frozen blocks of snow, pieces of sod, animal skins, or stone. The Sherpas of the Himalayas build their houses of stone, making very thick walls. They live on top of the yak stables so that the warmth given off by the animals helps warm the home.

Clothing Tundra dwellers wear layers of clothing during the cold weather. If traditional, these clothes are made from the skins of animals. Caribou skin is a favorite choice because it is very warm yet light. During winter, two pairs of pants are worn. The inside pair is made with the animal hair facing in for warmth. The outside pair is made with the hair facing outward to provide waterproofing.

People on the tundra may wear parkas, which are jackets with hoods. The lining of the hood is usually wolf or wolverine hair. Sealskin is desirable for boots because it is waterproof. Bird skin lines the boots to make them warm. Moss, which is very absorbent, may be dried out and used for baby diapers.

Economic factors For traditional hunter-gatherers, possessions are almost meaningless. They do not have traditional jobs or seek to gain wealth. They spend their time finding the food they need and making clothing, weapons, and tools. In the past, reindeer herders were totally self-sufficient, getting food, clothing, and shelter from their herds. After World War II (1939–1945) the economy of many tundra regions changed.

Many minerals have been discovered in Arctic tundra. Coal, iron, nickel, gold, tin, and aluminum are found in eastern European countries

The Wheels that Dug Lakes

The tundra environment is very fragile and its ability to restore itself is limited. When a construction truck or bulldozer drives on the tundra during the summer it leaves ruts in the ground. When the sun hits these ruts it causes the permafrost to melt, which then causes erosion, and the ruts get bigger. Each summer more of the exposed permafrost melts and eventually the rut turns into a gully. During World War II (1939–1945), heavy trucks were driven on the tundra, leaving large ruts behind. The areas of permafrost that melted because of these ruts have grown so large over the years that some of them are now lakes.

Minerals have been found in the tundra. Gold was mined from this tunnel around Salmon Glacier, Hyder, Alaska. IMAGE COPYRIGHT NATALIA BRATSLAVSKY, 2007. USED UNDER LICENSE FROM SHUTTERSTOCK.COM.

such as Russia. Lead and zinc are mined in Greenland and gold in Canada. The largest mining activity on the tundra is drilling for natural gas and oil. Oil found on tundra changed the economy of Western nations, who no longer need to rely so heavily on the Middle East for their oil supplies. The discovery of oil and the need to build a pipeline to move it brought many non-native people to the Alaskan tundra. The creation of jobs was a boost to the economy of the countries that provided the work force.

After World War II, the Arctic tundra was the site of military activity. In North America, the United States built a series of radar stations from Alaska to Greenland called the Distant Early Warning (DEW) Line. This brought new residents and new jobs to the area.

Impact of human life on the tundra   Due to its relative isolation, either in the far north or high in the mountains, the tundra biome has not been affected as much by the presence of humans as have many other biomes. Its harsh environment has helped keep people away, but this is changing.

Use of plants and animals   When native peoples use plants and animals only for their own food and necessary materials, wildlife populations remain stable. When people hunt with guns for commercial reasons, more animals are killed than are born. In the 1800s and throughout much of the 1900s, tundra animals were overhunted. Musk oxen were almost wiped out when their meat was sold to sailors on whaling ships. The skins were traded and baby musk oxen were sold to zoos. Caribou were so severely overhunted that their herds were reduced by 90 percent.

Many countries have since set limits on the number of tundra animals that can be killed each year, including caribou, musk oxen, and polar bears.

Quality of the environment   The quality of the Arctic tundra environment has been threatened by the effects of mining, the use of pesticides, and air pollution.

Effects of mining   Mining and drilling operations pollute the air, lakes, and rivers, and damage the land. In Russia, the land around some nickel mines has become so polluted that all the plants have died. With no plant life to anchor it, the soil has washed away. Delicate plants are trampled as roads, airstrips, and houses are constructed. These plants take years to grow back, and some never recover. When mines are abandoned, the old equipment and debris may be left behind, becoming eyesores.

Animals will not go near mining and drilling operations because of the noise and activity. This may cut them off from familiar food and water sources. The Alaskan pipeline, for example, was built across the caribou migration route. In order to correct this problem, the pipeline has been raised like an overpass so they can walk underneath, but the road beside the pipeline may be a problem. Scientists do not know what long-range effects these human-made additions will have on the animals.

The increasing number of people moving to the tundra to work in mining operations has created a need for houses, towns, and additional roads to bring in supplies. Getting rid of waste is a serious problem. The cold weather prevents decomposition of garbage and it cannot be buried in the frozen ground. The choices are to ship it out, which is very expensive, or create above-ground garbage dumps, which destroy the landscape.

The Aurora Borealis, or Northern Lights. IMAGE COPYRIGHT STEPHEN KIERS, 2007. USED UNDER LICENSE FROM SHUTTERSTOCK.COM.

Use of pesticides Pesticides (poisons) affect tundra wildlife. A migratory bird such as a goose may eat plants sprayed with pesticides in the United States, where it spends the winter. When it returns to the Arctic in summer, a predator such as a fox may kill it. The fox and her young eat the goose. A falcon or polar bear might eat one of the young foxes. In this way, the pesticide contaminates a number of animals.

Peregrine falcons and polar bears have been especially affected by pesticide use. Some pesticides make falcon eggs soft, causing them to break open before the chick has fully matured. Pesticides that have built up in the bodies of polar bears after they have eaten a number of contaminated animals can, ultimately, kill them.

Air pollution The quality of the environment is affected by what happens outside the tundra. Air pollution may travel to the tundra from

Alpine tundra in the Rocky Mountains generally begins forming from 4,000 to 10,000 feet (1,219 to 3,048 meters) above sea level. Since its elevation is fairly high, Niwot Ridge is characterized by low temperatures throughout the year. The annual mean temperature is 5°F (−3.7°C). Temperatures in January average about 8°F (−13°C) and in July about 47°F (8°C).

This tundra is surrounded by subalpine forest at the lower elevations. Where the forest and tundra meet, subalpine meadows and patches of krummholz (dwarf trees) exist. Other physical features include a cirque glacier, the Arikaree, which is a U-shaped glacier with its open end facing down the valley. Talus slopes formed by the accumulation of rock fragments are also found here.

Some areas are snow covered. Strong winds that occasionally reach 170 miles (273 kilometers) per hour blow snow from other areas, leaving them bare. Meltwater (water melted from ice or snow) is found only where there is snow. Although precipitation in summer is uneven, yellow, pink, and purple flowers color the tundra in warm months. Plants include snow buttercup, old-man-of-the-mountain, Parry's primrose, and shooting stars.

Many birds live on this tundra only in summer. These include the water pipit, horned lark, and white-crowned sparrow. The only year-round resident is the white-tailed ptarmigan.

Thirty-two species of mammals live on Niwot Ridge. Small plant-eating mammals include deer mice, voles, golden-mantled ground squirrels, pikas, yellow-bellied marmots, snowshoe hares, and porcupines. Badgers and weasels are seen from time to time.

Many scientists and students work in the area studying the weather, analyzing the soil, and observing plant and animal life. Niwot Ridge is designated as a Biosphere Reserve by the United Nations' Educational, Scientific, and Cultural Organization (UNESCO) and an Experimental Ecology Reserve by the United States Department of Agriculture (USDA) Forest Service.

The ridge is part of Roosevelt National Forest. When visitors are allowed on the tundra, they are encouraged to use the trails designated for this purpose because even one footprint can damage the fragile landscape.

Bering Tundra
Location: Seward
Peninsula, Alaska
Area: 23,400 square miles
(60,610 square kilometers)
Classification: Arctic
tundra

Bering Tundra   The Bering Tundra is a western extension of the arctic coastal plain, a broad lowland in western Alaska between Kotzebue and Norton Sound. It is located on the Seward Peninsula on the American side of the former Bering Land Bridge (a narrow strip of land with water on both sides that at one time connected two continents) and is usually in the form of snow.

The Bering Tundra has cold winters and cool summers. Although summer temperatures in a polar climate rarely exceed 50°F (10°C), a high of 90°F (32°C) in summer has been recorded here. In winter, the low has reached –70°F (–57°C). Annual precipitation averages 17 inches (43 centimeters).

Thousands of shallow lakes and marshes (wetlands with poorly drained soil and nonwoody plants) are found along the coast. Two large rivers, the lower Yukon and the Kuskokwim, flow out of the province to the southwest. The terrain on the peninsula varies from lava fields to hot springs to tundra.

Much of the tundra is less than 1,000 feet (305 meters) above sea level. Some small mountain groups range from 2,500 to 3,500 feet (762 to 1,067 meters) high. The highest point is Mount Osborn, which reaches a height of 4,714 feet (1,436 meters).

Permafrost lies under most of the area and the active layer of soil is considered young and undeveloped. Despite permafrost and a growing season that can be as short as two weeks, there is still a wide variety of vegetation.

Species of dwarf trees, like birch, border the tundra. Birch, willow, and alder thickets are found between shoreline and forest. The lower Yukon and Kuskokwim Valleys are dominated by white spruce and cottonwood. Typical tundra plants include sedges, lichens, mosses, and cottongrass tussocks. Labrador tea, cinquefoil, and brightly colored forbs also grow in the region.

The coast provides habitats for migrating waterfowl and shore birds. Other bird species include ospreys, falcons, grouse, ravens, golden eagles, and various hawks and owls.

Mammals living in the tundra include musk oxen, brown and black bears, wolves, wolverines, coyotes, and a large herd of moose. Snowshoe hares, red foxes, lynxes, beavers, and squirrels summer here. Polar bears, walruses, and arctic foxes are sometimes seen along the northern coast of the Bering Sea.

Caribou migrate every summer to the tundra to give birth to their calves. They sometimes mix with domestic caribou, causing the native Eskimo herders to lose many of their animals, which join the wild herds.

An Inupiat Eskimo settlement of about 170 people is located in Wales on the western tip of the Seward Peninsula. Fewer than 60 miles (96 kilometers) from Siberia, Wales is one of the Alaskan settlements closest to Russia. Here, and in other places on the peninsula, native hunter-gatherers still live off the land. They are faced with the challenge of adapting to changes caused by development, especially by the exploitation of natural resources such as oil.

During the early twentieth century, thousands of non-native people came to Alaska in search of gold. As a result, more than a million dollars in gold was taken from the peninsula.

Kola Peninsula   Murmansk is a province in northwestern Russia on the Kola Peninsula, a peak of high land that juts out into the ocean between the Barents and White Seas. Most of the peninsula lies across the Arctic Circle. It extends about 190 miles (305 kilometers) from north to south, and 250 miles (400 kilometers) from east to west. Elevations in the peninsula's Khibiny Mountains reach 3,907 feet (1,191 meters), and Arctic tundra covers the northern areas.

The largest town is the ice-free port of Murmansk, on the eastern shore, with a population of more than one million people. Fishing is the main occupation along the coast, but mining is the most important part of the economy.

Many minerals are found on the peninsula, including the world's largest deposits of apatite, a mineral rich in phosphorus and used for fertilizer production. Nephelinite (a source of aluminum), zirconium, iron, and nickel are also mined here.

Bogs (wetlands with wet, spongy, acidic soil) are widespread and form in areas where the soil is saturated by water. Since summer meltwater cannot drain into the permafrost and temperatures are too cool for evaporation, conditions are ideal for bog formation.

Despite problems with permafrost and thin, poorly developed topsoil, the peninsula has been called a botanical garden. Mosses, lichens, and dwarf Arctic birch cover most of the region. A forested area in the south has birch, spruce, and pine. Other typical tundra vegetation include lichens, Lapp rhododendrons, Arctic willows, and white mountain avens, a yellow-flowered member of the rose family.

Bird life typical to the tundra includes Siberian jays, Siberian titmice, grouse, and ptarmigans. Migrating seabirds include eider ducks, skuas, gulls, and Atlantic puffins. Lemmings, beavers, otters, and brown bears

Kola Peninsula
Location: Murmansk Oblast Province, northern Russia
Area: 40,000 square miles (100,000 square kilometers)
Classification: Arctic tundra

are common mammals, and thousands of reindeer migrate to the tundra in summer.

In the interior, a few thousand Sami are engaged in reindeer herding, which was the basis of their economy until the twentieth century. Families lived in tents and migrated with their herds. This way of life is disappearing as families now have permanent homes and only the herders move with the reindeer.

The deepest hole ever created by humankind is in the Kola Peninsula. The drilling experiment taken on by the Russians in 1962 through 1994, in order to investigate Earth's crust, ended with a hole 7 miles (12 kilometers) deep.

The western part of Kandalaksha Bay and the area south of Murmansk are wildlife reserves. No visitors are allowed in these areas. They are headquarters only for scientific research.

Northeastern Svalbard Nature Reserve The Svalbard Reserve is one of the largest and most important nature reserves in Norway. The area became protected by the Norwegian government in 1973 and includes North East Land, Kvit Island, Kong Karls Land, smaller adjoining islands, and the surrounding territorial waters. North East Land, the largest part of northeastern Svalbard, is covered by glaciers and ice caps all year long. *Svalbard* means "the land with the cold coast."

Arctic tundra covers the reserve, which is characterized by high winds, extremely cold temperatures, and permafrost that reaches depths of 1,640 feet (500 meters). Only the upper 6 to 10 feet (1.8 to 3 meters) of ground thaw in the summer, making it impossible for trees or plants with deep roots to grow.

In most Arctic climates winters are extremely severe and summers are short and cool. A branch of the warm ocean current called the North Atlantic Drift, moderates the climate here. Temperatures range from 59°F (15°C) in the summer to –40°F (–40°C) in the winter.

The growing season is very short, lasting only a few weeks in the summer. Plant life is typical for Arctic tundra and includes more than 150 species of plants. There are many lichens and mosses, and tiny polar willows and dwarf birches grow along the tundra borders.

Almost twenty species of birds nest on the islands. These include murres, which are diving shorebirds with stocky bodies, short tails, wings, and webbed feet. Several types of gulls, ptarmigans, arctic terns, and

Northeastern Svalbard Nature Reserve
Location: Svalbard Islands, Norway
Area: 7,350 square miles (19,030 square kilometers)
Classification: Arctic tundra

phalaropes also nest here. Two species of ptarmigan, the rock ptarmigan and the willow ptarmigan, are the only birds that live here year round.

There are fewer species of mammals than birds in the area. Polar bears, Arctic foxes, reindeer, and musk oxen live in the reserve. Polar bears in this region eat only meat; seals are a big part of their diet. Not native to the reserve, the musk ox was imported from Greenland in 1929.

Since trapping is an important economic activity, land game, such as foxes, reindeer, and bears, is protected by law so it does not become endangered by overhunting. Hunters once lived on the tundra, but very few spend the winter there. Most live in permanent communities at lower elevations. Svalbard's economy depends upon industrial operations and coal mining.

Tromsø University in Norway, in cooperation with the University of Alaska, runs a Northern Lights research station in the reserve.

Katmai National Park and Preserve   The Katmai National Park and Preserve is located on the northeast coast of the Alaska Peninsula, along Shelikof Strait. The northern part of the park is Alpine tundra. Moving south, the terrain changes to forests, which line the bays and fjords (narrow inlets or arms of the sea bordered by steep cliffs).

Katmai tundra begins at a relatively low elevation, around 2,000 to 2,300 feet (600 to 700 meters) above sea level. The highest peaks in the park reach 7,585 feet (2,312 meters).

The climate is cold and windy with harsh winters; short, cool, summers with constant winds. In Katmai, the average summer temperature is 60°F (15.5°C). In winter there are only about six hours of sunlight each day.

The terrain in this park includes glaciers, waterfalls, and mountains. The Valley of Ten Thousand Smokes, where thousands of steam vents once came through the valley floor, was formed in 1912 when the Novarupta Volcano erupted violently. Only a few active vents remain.

Forests of blue spruces, alders, and willows border the tundra. Higher up, heaths, mountain avens (a member of the rose family), and bearberries grow.

Katmai National Park and Preserve
Location: Alaska
Area: 5,806 square miles (15,038 square kilometers)
Classification: Alpine tundra

Mammals include brown bears, wolves, foxes, moose, and caribou. The park is home to the largest population of protected brown bears in the world. The Alaskan brown bear, also known as the grizzly bear, is the world's largest carnivore and feeds on red salmon that spawn in the area.

People do not live here permanently because it is too inhospitable, but many visitors come to camp.

Taymyr Taymyr (also spelled Taimyr or Tajmyr) is a province located in northeastern central Russia. Mostly Arctic tundra, it extends from the Taymyr Peninsula south to the northern edge of the Central Siberian Plateau. The area includes the Severnaya Zemlya archipelago (a group of many islands) in the high Arctic.

The climate in Taymyr is exceptionally severe, with prolonged, bitter winters. The average temperature in January is $-22°$F $(-30°$C) and in July from $35°$ to $55°$F $(2°$ to $13°$C). There are few sunny days. Precipitation ranges from 4 to 14 inches (11 to 35 centimeters) annually.

The region has a variety of landforms, including mountains, lowlands, and plateaus (raised, flat-surfaced areas). The mountains of Byrrang are at an elevation of 3,760 feet (1,146 meters). In the Arctic, plateaus of ice can be as thick as 4,900 feet (1,500 meters). This ice was formed when glaciers covered much of the Arctic more than 10,000 years ago. The Plateau of Plutoran has an elevation of 4,399 feet (1,341 meters).

A variety of minerals are found in Taymyr, including complex ores such as copper, nickel, platinum, gold, and coal. Gas is also mined.

The soil in Taymyr is typical tundra soil. The active layer of topsoil above the permafrost thaws in summer and freezes in winter. Bogs are widespread and form in areas where the soil is saturated with water. Since summer meltwater cannot drain into the permafrost and temperatures are too cool for evaporation, conditions are ideal for bog formation.

Although the growing season is only a few weeks long, plants take advantage of the long hours of daylight during the short summer months. Most of the area is covered with mosses, sedges, rushes, and some grasses. Lichens grow on the hillsides and bilberries, a type of blueberry, grow in clusters. Many flowering plants provide color. On the edge of the tundra, dwarf willows and birch trees grow.

Taymyr is the summer home for the red-breasted goose, which is a threatened species. Steller's eider, a large sea duck, is found on the peninsula. The mammal population includes reindeer, sable, wolf, fox, and white hare. Polar bears live farther north.

Few people live above the Arctic circle. Those that do include Russians, Dolgans, Nenets, Ukrainians, and Nganasans. Their chief economic activities are hunting, reindeer herding, fishing, and fur farming. Animals whose fur is used for clothing, such as the sable, are raised domestically

Taymyr
Location: North Central Russia
Area: 332,850 square miles (862,100 square kilometers)
Classification: Arctic tundra

rather than trapped in the wild. Only two cities are located in Taymyr province, the capital, Dudinka, and the port of Dikson. Their combined population is estimated at about 55,000.

Many traditional native lifestyles are slowly dying out. The Dolgan, for example, were primarily reindeer herders and lived a nomadic lifestyle for the last several hundred years. Since the 1970s their lifestyle has become less nomadic. Now they rely on gardening as well as hunting for food.

The Russian government established the Great Arctic Reserve in 1993 on the Arctic tundra for shorebird conservation and research. This area is the breeding zone for many shorebirds who summer in Africa or along the Atlantic coasts of Europe.

Gaspé   The Gaspé (Gas-PAY) Peninsula lies in eastern Quebec, Canada. The forests and alpine tundra extend east-northeastward for 150 miles (240 kilometers) from the Matapédia River into the Gulf of St. Lawrence. The name Gaspé is from the Mi'kmaq Indian word *gespeg*, which means "land's end."

The tundra covers 24 percent of Quebec and lies in the Chic-Choc, or Shickshock, Mountains, which are part of the Appalachian range. They are the highest in Quebec, with Mount Jacques Cartier rising to 4,160 feet (1,268 meters). The tundra on the Gaspé Peninsula begins around 3,280 feet (1,000 meters) above sea level.

Gaspé lies between a humid continental climate and a subarctic climate. Continental climates are warmed somewhat in summer as tropical air moves north. In winter the cold polar air moves south. Gaspé faces severe winters, which have kept the population sparse. Temperatures can range from lows of –11° to 14°F (–10° to –24°C) in January to highs of 52° to 68°F (11° to 20°C) in July.

The land is crisscrossed by a number of rivers, and the peninsula is surrounded on three sides by the St. Lawrence River and the Gulf of St. Lawrence. A coniferous (evergreen) forest separates the province's deciduous (trees that loose their leaves at the end of each growing season) forests to the south from the subarctic and tundra areas where no trees grow.

Arctic-alpine plants that grow along the cliffs of the peninsula include lichens, mosses, sedges, grasses, and other low woody and leafy plants. On the edges of the tundra maples, yellow birches, white spruces, and balsam firs grow.

Gaspé
Location: Eastern Quebec, Canada
Area: 11,390 square miles (29,500 square kilometers)
Classification: Alpine tundra

Many seabirds are found along the cliffs, including black-legged kittiwakes, double-crested cormorants, black guillemots, and razorbills. This is the only area of Quebec where caribou, moose, and white-tailed deer are found. Other mammals include foxes, lynxes, black bears, beavers, Arctic hares, and porcupines.

People do not live in this alpine tundra. It is a place for visitors and research only. The rest of the peninsula is sparsely populated, with about one-fifth of the people earning a living through agriculture. In summer months, many tourists visit the peninsula and support the economy. Lumbering, mining, and fishing are the major economic activities. Minerals mined in the area are copper, lead, and zinc.

The Mi'kmaq (also spelled Micmac) people have inhabited the Gaspé Peninsula for thousands of years. They were one of eight major tribes who comprised the Woodland First Nations peoples. Traditionally, they were a seasonally nomadic people, hunting moose, caribou, and small game in winter and fishing in summer. Their winter homes were conical (cone-shape) wigwams covered with birch bark or animal skins. In summer, they lived in open-air, oblong wigwams. Homes were portable and easy to put up or move. Travel was by canoe, toboggan, or snowshoe. Some of the Mi'kmaq still live on Gaspé, but many have migrated to the United States.

The Gaspésian Provincial Park is a large conservation area covering much of the peninsula. Another park, Forillon National Park, covers 93 square miles (240 square kilometers) at the northeastern tip of the peninsula.

Arctic National Wildlife Refuge   The Arctic National Wildlife Refuge in Alaska is one of the largest wildlife refuges in the world. To the north is the Arctic Ocean and to the south is Porcupine River. The refuge extends east to west more than 200 miles (322 kilometers) from the Trans-Alaska pipeline corridor to Canada, and almost 200 miles north to south from the Beaufort Sea to the Yukon Flats National Wildlife Refuge. It supports both Arctic and alpine tundra.

The Brooks Range, with peaks as high as 9,000 feet (2,743 meters), extends from east to west through the refuge. The Brooks Range is the highest mountain range within the Arctic Circle and is the northernmost extension of the Rocky Mountains in northern Alaska.

Winter on the refuge is long, and snow usually covers the ground at least nine months of the year. The average winter temperature here is

Arctic National Wildlife Refuge
Location: Alaska
Area: 19,000,000 acres
(7,600,000 hectares)
Classification: Arctic and alpine tundra

below 0°F (below –17°C). The wind chill factor makes the temperature feel like –100°F (–73°C) to exposed skin. In summer, the temperature averages 50°F (10°C). Annual precipitation ranges from 4 to 16 inches (10 to 40 centimeters).

Besides two tundra zones, the refuge includes barren mountains, forests, shrub thickets, and wetlands. The alpine tundra is crisscrossed by braided rivers and streams, with clusters of shallow, freshwater lakes and marshes.

The Arctic National Wildlife Refuge contains the greatest variety of plant and animal life of any conservation area within the Arctic Circle. Even though the growing season can be as short as two weeks, the continuous summer daylight helps plants grow rapidly. Reindeer moss, sedges, and flowering plants are common. Spruces, poplars, birches, and willow trees grow in areas surrounding the tundra.

More than 180 species of birds have been observed here. Peregrine falcons are common in the refuge as are rock ptarmigans, grebes, snow geese, plovers, and sandpipers.

Thirty-six species of land mammals live here, including all three species of North American bears (black, brown, and polar). Smaller mammals include lynxes, wolverines, lemmings, Arctic hares, and wolves.

In the summer, tens of thousands of caribou give birth to and raise their calves on the tundra. These caribou migrate south in winter to northeastern Alaska and the northern Yukon, almost 1,000 miles (1,609 kilometers), farther than any other land animal. Even though their winter habitat is milder than their tundra breeding grounds, they still face Arctic weather. The caribou can dig through snow almost 2 feet (60 centimeters) deep to reach food, which consists of lichens, sedges, and grasses. The refuge protects most of the calving grounds for the Porcupine caribou herd, the second largest herd in Alaska, which numbers about 180,000 animals. It also protects a large part of their migration routes.

Dall sheep live in the refuge all year, mostly in the tundra. They stay close to rocky outcrops and cliffs where they are safe from predators that include wolves, golden eagles, bears, and humans. Dall sheep eat grasses, sedges, broad-leaved plants, and dwarf willows. In winter, when these foods are scarce, they eat lichens.

The area is home to several groups of native peoples and is the site of controversy over oil development. Kaktovik, an Inupiat Eskimo village, borders the refuge on the north side. The Inupiat have incorporated and, along with the government, own part of Alaska. The Gwich'in, another native group, are one of the few who have not incorporated and choose to maintain their reservation status and a subsistence lifestyle in their nine villages.

The Inupiat see the value of oil development in the area. It provides jobs and supports schools and community improvements. The Gwich'in see it as endangering the fragile tundra environment and disturbing migration routes and breeding places for the caribou.

Huascaran National Park   Huascaran (hwa-SCAH-run) National Park is a place of biological diversity. Situated in the Andes Mountains in Peru, the landscape is dominated by mountain peaks, glaciers, lakes, and alpine tundra. It became a national park in 1975 and a World Heritage Site in 1985. The Quechua (QWAY-chwah) people live near the park and maintain a traditional lifestyle.

Alpine tundra, characterized by cold winds, is found on mountains above the timberline. The lower portion of the mountain is covered by tropical rain forest.

The snowcapped peak of Mount Huascaran, the highest mountain in Peru, rises to an elevation of 22,205 feet (6,768 meters) above sea level. The climate varies, with temperatures getting colder and rainfall decreasing as the elevation increases. The tundra climate has an average annual temperature below 50°F (10°C).

Much of Huascaran National Park is covered with permanent ice and snow. A permanent snowline begins between 14,700 and 16,400 feet (4,480 and 5,000 meters). Just below the tundra is the *puna*, an area of alpine grasslands with no trees. Below that is the *tierra fria*, the cool land, where most of the people live. This area has forest and land suitable for growing wheat and potatoes.

Vegetation is typical of alpine tundra in other parts of the world. Mosses and lichens grow on bare rocks. Short flowering plants and grasses, cushion plants, and sedges are common.

The South American, or Andean, condor nests on the cliffs in the high tundra. A member of the vulture family, it is one of the largest flying birds in the world with a wing span of about 10 feet (3 meters). The condor is carnivorous and feeds on dead animals.

Mammals in the Andes are different from those in other alpine tundra. The llama, alpaca, vicuña, and guanaco are native to this part of the world. Although they are all relatives of the camel, they have no hump. The vicuña is an endangered species hunted for its fleece, from

Huascaran National Park
Location: Peru
Area: 840,000 acres
(340,000 hectares)
Classification: Alpine tundra

which some of the world's finest wool is made. The chinchilla, a small rodent common here, was hunted almost to extinction for its fur.

South American Indians living near the Huascaran Park are the Quechua and the Aymara, who have adapted to cold, high-altitude living. They have larger lungs and hearts, and their blood contains more red blood cells, allowing it to hold more oxygen. This makes it easier for them to live and work in the thin atmosphere of the mountains. They are also able to walk barefoot on icy rocks. These native peoples lead isolated lives as herders and farmers. Women make clothes and blankets to sell at market, spinning their own wool and weaving fabrics.

For More Information

BOOKS

Cox, C.B., and P.D. Moore. *Biogeography and Ecological and Evolutionary Approach.* 7th ed. Oxford: Blackwell Publishing, 2005.

Gaston, K.J., and J.I. Spicer. *Biodiversity: An Introduction.* 2nd ed. Oxford: Blackwell Publishing, 2004.

Grzimek, Bernhard. *Grizmek's Animal Encyclopedia.* 2nd edition. Volumes 4-5. *Reptiles,* edited by Michael Hutchins, Dennis A. Thorney, Paul V. Loiselle, and Neil Schlager. Farmington Hills, MI: Gale Group, 2003.

Houghton, J. *Global Warming: The Complete Briefing.* 3rd ed. Cambridge: Cambridge University Press, 2004.

McGonigal, D., and L. Woodworth. *Antarctica: The Complete Story.* London: Frances Lincoln, 2003.

Moore, P.D. *Biomes of the Earth: Tundra.* New York: Chelsea House, 2006.

PERIODICALS

Payette, Serge, Marie-Josee Fortin, and Isabelle Gamache. "The Subarctic Forest—Tundra: The Structure of a Biome in a Changing Climate." *BioScience.* 51. 9 September 2001: 709.

Pennisi, Elizabeth. "Neither Cold Nor Snow Stops Tundra Fungi." *Science.* 301. 5,638 September 5, 2003: 1,307.

Weir, Kirsten L. "Don't Tread on It." *Natural History.* 110. 9 November 2001: 37.

ORGANIZATIONS

Center for Environmental Education, Center for Marine Conservation, 1725 De Sales St. NW, Suite 500, Washington, DC 20036

Environmental Defense Fund, 257 Park Ave. South, New York, NY 10010, Phone: 800-684-3322; Fax: 212-505-2375, Internet: http://www.edf.org.

Environmental Network, 4618 Henry Street, Pittsburgh, PA 15213, Internet: http://www.envirolink.org.

Environmental Protection Agency, 401 M Street, SW, Washington, DC 20460, Phone: 202-260-2090, Internet: http://www.epa.gov.

Friends of the Earth, 1025 Vermont Ave. NW, Ste. 300, Washington, DC 20003, Phone: 202-783-7400; Fax: 202-783-0444.

Greenpeace USA, 1436 U Street NW, Washington, DC 20009, Phone: 202-462-1177; Fax: 202-462-4507, Internet: http://www.greenpeaceusa.org.

Nature Conservancy, 1815 N. Lynn Street, Arlington, VA 22209, Phone: 703-841-5300; Fax: 703-841-1283, Internet: http://www.tnc.org.

Sierra Club, 85 2nd Street, 2nd Floor, San Francisco, CA 94105, Phone: 415-977-5500; Fax: 415-977-5799, Internet: http://www.sierraclub.org.

World Wildlife Fund, 1250 24th Street NW, Washington, DC 20037, Phone: 202-293-4800; Fax: 202-229-9211, Internet: http://www.wwf.org.

WEB SITES

Arctic Studies Center: http://www.nmnh.si.edu/arctic (accessed August 9, 2007).

Long Term Ecological Research Network: http://lternet.edu (accessed August 9, 2007).

National Geographic: http://www.nationalgeographic.com (accessed August 9, 2007).

National Park Service, Katmai National Park and Preserve. http://www.nps.gov/katm/ (accessed August 9, 2007).

National Science Foundation: http://www.nsf.gov (accessed August 9, 2007).

Thurston High School: Biomes: http://ths.sps.lane.edu/biomes/index1.html (accessed August 9, 2007).

"The Tundra Biome." University of California, Museum of Palentology. http://www.ucmpberkeley.edu/exhibits/biomes/tundra.php (accessed August 9, 2007).

Wetland

Wetlands are areas covered or soaked by ground or surface water often enough and long enough to support special types of plants that have adapted for life under such conditions. Wetlands occur where the water table (the level of groundwater) is at or near the surface of the land, or where the land is covered by shallow surface water (usually no deeper than about 6 feet [1.8 meters]). They form the area between places that are always wet, such as ponds, and places that are always dry, like forests and grasslands.

Many wetlands are not constantly wet, experiencing what is called a wet/dry cycle. Some are wet for only part of the year, like those that are drenched during heavy seasonal rains. Some have no standing water but, because they are near the water table, their soil remains saturated (soaked with water). Others may dry out completely for long periods, sometimes years.

Wetlands are among the world's most productive environments with high biodiversity (a large variety of life forms). Only rain forests and coral reefs have more biodiversity.

Wetlands are found all over the world, in every climate from the frozen landscape of Alaska to the hot zones near the equator, except Antarctica. In the continental United States, there are about 110,000,000 acres (44,500,000 hectares) of natural wetland, and some types are found in every state. Alaska contains about 175,000,000 acres (71,000,000 hectares) of wetlands.

How Wetlands Develop

Wetlands have a life cycle that begins with their formation and may involve many changes over time.

Formation   Many wetlands were formed when glaciers retreated after the last ice age, about 10,000 years ago. Some of the glaciers left depressions

The Great egret makes its home in the marshy areas of wetlands where its long legs are ideal for wading in the shallow waters.
IMAGE COPYRIGHT GLEN JONES, 2007. USED UNDER LICENSE FROM SHUTTERSTOCK.COM.

Prairie potholes are small marshes found by the millions throughout the north central area of the United States and south central Canada. Usually no more than a few feet deep, they may cover as much as 10,000 acres (4,000 hectares). Prairie potholes are important to migratory waterfowl as places to rest, feed, and nest. In Africa, these small marshes are called dambos. In North America it is estimated that two-thirds of all waterfowl are hatched in prairie potholes.

Wet meadows are freshwater marshes that frequently become dry. They look like grasslands but the soil is saturated. They are common in temperate and tropical regions around the world, including the midwestern and southeastern United States. Their dominant plants are sedges, flowering plants that resemble grass.

Riparian wetlands are marshes found along rivers and streams. They range in width from a few feet (meters) to as much as 12 miles (20 kilometers). Smaller riparian wetlands are common in the western United States. Larger ones are located along large rivers, such as the Amazon in South America. Riparian wetlands are unique from others in that the vegetation and life forms found immediately adjacent to the river or streams differ from those of the surrounding area, usually a forest. This marked contrast produces a diversity and enhances the benefits for wildlife, both in food (grasses and sedges) and in shelter (shrubs and trees) in a relatively small place.

Mangrove tree, here seen at low tide. IMAGE COPYRIGHT ECOPRINT, 2007. USED UNDER LICENSE FROM SHUTTERSTOCK.COM.

A wash is a dry streambed that becomes a wetland only after a rain. Washes are found in dry plains and deserts. The plant life they support usually has a short growing season and disappears during dry periods. In North Africa and Saudi Arabia washes are called *wadis.* In the American west they are called *arroyos.*

Saltwater marsh Saltwater marshes, also called salt marshes, are found in low, flat, poorly drained coastal areas. They are often flooded by salt water or brackish water (a mixture of both fresh and salt water). Saltwater marshes are especially common in deltas, along low seacoasts, and in estuaries (arms or inlets of the sea where the salty tide washes in and meets the freshwater current of a river). They can be found in New Zealand, in the Arctic, and along the Atlantic, Pacific, Alaskan, and Gulf coasts.

Saltwater marshes are greatly affected by tides, which raise and lower the water level on a daily basis. A saltwater marsh may have tidal creeks, tidal pools, and mud flats, each of which has its own ecosystem (a network of organisms that have adapted to a particular environment).

The high salt content of the sea water in the marsh makes it hard for plants to adapt. Grasses such as marsh grass, cord grass, salt hay grass, and the grasslike needlerush thrive here.

Peatland Peatlands are wetlands in which peat has formed. Peat is a type of soil made up of the partially decayed remains of dead plants, such as

Cranberries grow in bogs in the cooler parts of the northern hemisphere. IMAGE COPYRIGHT LIJUAN GUO, 2007. USED UNDER LICENSE FROM SHUTTERSTOCK.COM.

sphagnum moss and even trees. In peatlands, dead plant matter is produced and deposited at a greater rate than it decomposes. Over time, sometimes over thousands of years, a layer of this plant matter is formed that may be as deep as 40 feet (12 meters). Several conditions are necessary for peat to form. The soil must be acidic, waterlogged from frequent rains, and low in oxygen and nutrients. The bacteria responsible for decomposing plant matter cannot thrive under these conditions, resulting in accumulation of the layers of partially decayed plant matter.

Peatlands cover three percent of Earth's land and freshwater surface area. Peatlands are located in colder, northern climates and in the tropics. They can be found in Russia (mainly Siberia), China, Scandinavia, northern Europe, England, Ireland, Canada, and the northern United States.

The types of peatlands are temperate bogs, fens, and tropical tree bogs.

Temperate bog   Ninety percent of peatlands are found in northern temperate climates where they are called bogs. Bogs have a soft, spongy, acidic soil that retains moisture, which comes primarily from rainwater. Some bogs have taken as long as 9,000 years to develop and are sources of peat that can be burned as fuel.

Temperate bogs have been described as soft, floating carpets. The carpet is made mostly of sphagnum moss, which can be red, orange, brown, or green and may grow as long as 12 inches (30 centimeters). Sphagnum moss can grow in an acidic environment. It holds fresh rainwater in which other bog plants can grow. The high concentration of sphagnum moss has led other bog plants to develop unusual adaptations for obtaining nutrients. For example, the bog myrtle forms a partnership with the bacteria in its roots to get extra nitrogen. Other plants that grow in bogs include grasses, small shrubs such as leatherleaf, flowering plants such as heather, and poison sumac. Some rare wildflowers, such as the lady slipper orchid and the Venus flytrap, are found in bogs.

Most bogs lie in depressed areas of ground. Some, called raised bogs, grow upward and are about 10 feet (3 meters) higher than the surrounding area.

Blanket bogs are shallow and spread out like a blanket. They form in areas with relatively high levels of annual rainfall. Their average depth is 8 feet (2.6 meters). Blanket bogs are found mainly on lowlands in western Ireland and in mountain areas. After a heavy rainfall, bogs located on steep slopes can wash down like a huge landslide of jelly and cover cattle, farms, and even villages.

> ### Nature's Diapers
>
> Sphagnum moss, which is found in bogs, can absorb many times its weight in water. At one time, certain Native Americans dried this moss and used it as diapers for their babies.

Fen   When a peatland is less acidic and is fed by mineral-rich groundwater, it is called a fen. Plants characteristic of fens are grasses, sedges, and reeds rather than sphagnum moss, and the soil does not become as acidic as in a bog. Less peat accumulates in a fen and it becomes only about 6 feet (2 meters) thick. As plant matter accumulates in a fen over time, it may form a raised bog.

Tropical tree bog   Bogs found in tropical climates are called tree bogs. These bogs produce peat from decaying trees rather than from sphagnum moss. The trees most commonly found in these bogs are broad-leaved evergreens. Temperatures are warmer than in temperate regions, causing decay to occur more rapidly. As a result, not as much peat develops. The only source of water is rain.

Tropical tree bogs can be found in South America, Malaysia, Africa, and Indonesia.

Climate

Unlike some other biomes, wetlands do not have a characteristic climate. They exist in polar, temperate, and tropical zones, but not usually in deserts. They are very sensitive to changes in climate, such as a decrease in precipitation (rain, sleet, or snow). The amount of precipitation and changes in temperature affect the growth rate of wetland plants. Some wetlands are seasonal, which means they are dry for one or more seasons of the year.

Temperature   Temperatures vary greatly depending on the location of the wetland. Many of the world's wetlands are in temperate zones (midway between the North and South Poles and the equator). In these zones, summers are warm and winters are cold, but temperatures are not extreme. Wetlands found in the tropic zone, which is around the equator, are always warm. Temperatures in wetlands on the Arabian

Peninsula, for example, can reach 122°F (50°C). In northeastern Siberia, which has a polar climate, wetland temperatures can be as cold as –60°F (–51°C).

Rainfall   The amount of rainfall a wetland receives depends upon its location. The average rainfall for wetlands in Wales, Scotland, and western Ireland is about 59 inches (150 centimeters) per year. Wetlands in Southeast Asia, where heavy rains occur, can receive up to 200 inches (500 centimeters). In the northern areas of North America, wetlands exist where as little as 6 inches (15 centimeters) of rain fall each year.

Geography of Wetlands

The geography of wetlands involves landforms, elevation, and soil.

Landforms   Landforms found in wetlands depend upon location, soil characteristics, weather, water chemistry, dominant plants, and human interference. Their physical features are often short-lived, especially if they are near floodplains or rivers, which can cause abrupt changes. Wetlands usually form in a basin where the ground is depressed, or along rivers and the edges of lakes.

A view of the wetlands and woods near Ocean City in Maryland. IMAGE COPYRIGHT ANDREW F. KAZMIERSKI, 2007. USED UNDER LICENSE FROM SHUTTERSTOCK.COM.

Elevation   Wetlands are found at many elevations (the height of an area in relation to sea level). Some wetlands in the Rocky Mountains in North America, for example, are at an elevation of 10,000 feet (3,048 meters).

Elevation is used to help classify some wetlands in Ireland. Bogs that are less than 656 feet (200 meters) above sea level are called Atlantic blanket bogs. Those that are more than 656 feet (200 meters) above sea level are called mountain blanket bogs.

Soil   An important characteristic of a wetland is its soil. Soil composition helps to determine the type of wetland and what plants and animals can survive in it. Almost all wetland soils are at least periodically saturated.

Wetland soils are hydric. This means they contain a lot of water but little oxygen. Only plants that that can adapt to these wet soils live in wetlands. The nutrients in the soil often depend upon the water supply. If the water source is primarily rain, the wetland soils do not receive as many minerals as those fed by groundwater. Soil in floodplains is very rich and full of nutrients, including potassium, magnesium, calcium, and phosphorus.

In some bogs associated with forests, decaying plant matter fully decomposes and is combined with sediments to form muck. This type of soil is dark and sticky. To be classified as muck, soil must contain not less than 20 percent organic (derived from living organisms) matter.

Plant Life

An important characteristic of any biome is its plant life or lack of plant life. More than 5,000 species of plants live in or near wetlands. Wetlands have high biological productivity (the rate at which life forms grow in a certain period of time). The higher its plant productivity, the more animal life a wetland can support. The kinds of plants found in a wetland are determined by several factors, especially the type of soil and the quantity of water.

Some plants, called hydrophytes, grow only in water or extremely wet soil. Sedges are an example. Mesophytes, such as reeds, need moist but not saturated soil. When a wetland dries up,

Pitcher plants are carnivorous. An insect crawls into the "pitcher" and cannot crawl back up. The plant digests the insect. IMAGE COPYRIGHT CHERYL CASEY, 2007. USED UNDER LICENSE FROM SHUTTERSTOCK.COM.

A Useful Nuisance

People often grumble about the water hyacinth because it multiplies and spreads quickly in open water. The plants become so thick that boats have a hard time moving through them. The water hyacinth has an important use, though. It absorbs and neutralizes many pollutants that would otherwise contaminate the water. It also has a high nutrient content, which makes it a good fertilizer.

the area may fill with plants called xerophytes. These are plants adapted to life in dry habitats and can survive where other wetland plants would wilt.

A submergent plant grows beneath the water and is found in deep marshes and ponds. Even its leaves are below the surface. Submergents include milfoil, pondweed, and bladderwort, an insect-eating plant.

Found in deep marshes, floating aquatics float on the water's surface. Some, like duckweed, have free-floating roots. Others, including water lilies, water lettuce, and water hyacinths, have leaves that float on the surface, stems that are underwater, and roots that are anchored to the bottom.

An emergent plant grows partly in and partly out of the water. The roots are usually under water, but the stems and leaves are at least partially exposed to air. They have narrow, broad leaves, and some produce flowers. Emergents include reeds, rushes, grasses, cattails, and water plantain.

Algae, fungi, and lichens   It is generally recognized that algae (AL-jee), fungi (FUHN-ji), and lichens (LY-kens) do not fit neatly into the plant category.

Algae   Most algae are one-celled organisms too small to be seen by the naked eye. They make their food by a process called photosynthesis (foh-toh-SIHN-thuh-sihs). Photosynthesis is the process by which plants use the energy from sunlight to change water and carbon dioxide from the air into the sugars and starches they need. Some wetland algae drift on the surface of the water, forming a kind of scum. Others attach themselves to weeds or stones. Some grow on the shells of turtles or inside plants or animals. Microscopic algae that can be found in saltwater marshes include diatoms and green flagellates (FLAJ-uh-lates). Desmids are a type of green algae found in bogs.

Fungi   Fungi are plantlike organisms that cannot make their own food by means of photosynthesis; instead, they grow on decaying organic matter or live as parasites (organisms that depend upon another organism for food or other needs).

Fungi grow best in a damp environment, which makes wetlands a favorable home. Common wetland fungi include mushrooms, rusts, and puffballs.

Lichens   Lichens are combinations of algae and fungi. The alga produces food for both itself and the fungi by means of photosynthesis. It is believed the fungus absorbs moisture from the air and provides shade. One of the most common wetland lichens is reindeer moss, an important food source for northern animals such as caribou.

Green plants   Most green plants have several basic requirements: light, air, water, warmth, and nutrients. In a wetland, light and water are in plentiful supply. Nutrients, primarily nitrogen, phosphorus, and potassium, are obtained from the soil. Some wetland soils are lower in these nutrients and low in oxygen. As a result, many wetland plants have special tissues with air pockets that help them survive.

Trees that grow in swamps, such as the mangrove and bald cypress, have shallow roots. Due to the lack of oxygen in the soil, the roots remain near the surface. Other trees found in mild climates include willows and alders. Palms are found in warm climates.

Wetland plants are classified as submergents, floating aquatics, or emergents, according to their relationship with water.

Common green plants   Common wetland green plants include mangrove trees, insectivorous plants, reeds, rushes, and sedges.

Trees that live in wetlands must tolerate having their roots wet for a long period of time. One of the best examples of this is the mangrove tree, which grows in tropical and subtropical saltwater swamps. Mangroves have done so well adapting to wetland conditions that there are more than 34,000,000 acres (14,000,000 hectares) of them in the world. One of the largest mangrove forests is the Sundarban Forest in Bangladesh. Mangroves can also be found in southern Florida and other tropical and subtropical areas.

The Hat Thrower

An unusual fungus called a hat thrower grows on dung (animal waste) deposited in wetlands. The fungus forms a tiny black bulb that looks like a hat. The hat explodes under pressure and pieces get thrown outward from the fungus and attach to other plants. When an animal eats those plants, the hat thrower passes through its digestive system. When the animal defecates, a new fungus grows on the pile of dung.

Water lilies and wild grasses grow are common wetland plants. IMAGE COPYRIGHT DANISH KHAN, 2007. USED UNDER LICENSE FROM SHUTTERSTOCK.COM.

A heron walks the marshy water of the Bolsa Chica Wetlands in Huntington Beach, California. AP IMAGES.

stabbing prey such as frogs and fish. Other bills, like those of the spoonbill, are designed to root through mud in search of food.

Wetlands provide a variety of hiding places where birds can build nests and protect their young. Dense underbrush or hollows in the ground are good hiding places, especially when nests are made from reeds, flowers, and grasses, which help them blend into their surroundings. Wood ducks build their nests in hollow cavities in trees lining the wetland shores.

Common birds Birds found in wetland environments can be grouped as wading birds, shorebirds, waterfowl, and perching birds.

Wading birds Wading birds, such as herons and egrets, have long legs for wading through the shallow water. They have wide feet, long necks, and long bills that are used for nabbing fish, snakes, and other food. Herons and egrets are the most common in freshwater marshes of North America. The great blue heron stands 4 feet (1.2 meters) tall. This is the tallest recorded wading bird in North America.

Shorebirds Shorebirds feed or nest along the banks of wetlands and prefer shallow water. Their feet are adapted for moving in water, and some have long, widely spread toes to prevent them from sinking in the mud.

The bills of shorebirds are designed to help them find food. The ruddy turnstone, for example, has a short, flattened, upturned bill, which helps it sift through mud or overturn pebbles and shells. A favorite food of the oystercatcher is the mussel. Each young oystercatcher learns from its parents a technique for opening the mussel shells to get at the meat. Some birds hammer a hole in the shell with their bills, and others prop the shells at an angle so they can pry them open. It may take the birds several years of practice to get the technique just right.

There are more than 200 species of shorebirds dependent on wetland habitats. These include sandpipers, which are found in marshes, wet woodlands, and on inland ponds, lakes, and rivers.

Waterfowl   Waterfowl are birds that spend most of the time on water, such as ducks, geese, and swans. Their legs are closer to the rear of their bodies than those of most birds, which is good for swimming, but awkward for walking. Their bills are designed for grabbing wetland vegetation, such as sedges and grasses, on which they feed.

Perching birds   Perching birds can be found living along wetland areas where food and shelter are readily available. They are land birds, with feet designed for perching. Their feet usually have three long toes in the front and one in the back. The barred owl is commonly found in swamps, while red-winged blackbirds live in cattail marshes. Bogs are home to golden plovers, skylarks, and meadow pipits.

Food   Plants and small animals in wetlands provide a ready food source for birds. Some birds feed on vegetation, while others are predatory. The marsh harrier, for example, feeds on mussels, small fish, or insects and their larvae. Herons and egrets feed on fish, frogs, and snakes.

Reproduction   All birds reproduce by laying eggs. Most male birds are brightly colored and sing to attract the attention of females. After mating, female birds lay their eggs in nests made out of many different materials. These nests may be found in a variety of places throughout the wetland area. Different species of birds lay varying numbers of eggs. The tufted duck, for example, lays six to fourteen eggs that hatch in less than a month. The ducklings begin to swim within days.

Mammals   Mammals are warm-blooded vertebrates that are covered with at least some hair and bear live young nursed with milk. Aquatic mammals, such as muskrats, have waterproof fur that helps them blend into their surroundings and webbed toes for better swimming. Some of these mammals live permanently in wetlands and others, such as raccoons, visit for food, water, and shelter during part of their lives.

Common mammals   American beavers are well-adapted to the wetland environment. They have webbed feet for powerful swimming

The Duck that Saved Wetlands

The Federal Duck Stamp Program began in 1934 when the U.S. Congress passed the Migratory Bird Hunting Stamp Act. The act required waterfowl hunters to purchase and carry Federal Duck Stamps that are attached to a hunting license or other document. The money was used to buy or lease waterfowl habitats from their owners and designate them as restricted areas. This helped save millions of acres of wetlands.

Over the years, efforts have also included saving wetland species that were declining in numbers, such as wood ducks, canvasback ducks, and pintail ducks. The program helps endangered species that rely on wetlands for food and shelter.

Anyone can own Federal Duck Stamps. State, international, and junior stamps are available. A duck stamp provides free admission to all National Wildlife Refuges where entrance fees are charged. To learn more about the duck stamps and to see pictures of them, visit http://www.nationalwildlife.com.

Wetlands are ideal for water birds such as the Black-bellied whistling duck. IMAGE COPYRIGHT FLORIDASTOCK, 2007. USED UNDER LICENSE FROM SHUTTERSTOCK.COM.

and warm, waterproof fur. Other mammals found in wetlands include the mink, the water shrew, and the Australian platypus. The sitatunga, a type of antelope, lives near swamps in central and east Africa. It feeds on emergent wetland plants.

Red deer Although peatlands generally do not support many species of animals, red deer live in the bogs. At 5 feet (1.5 meters) tall, it is the largest land mammal found in Ireland. It can be seen rolling in the peat in order to get rid of parasites and insects.

Muskrat Muskrats resemble beavers. They are heavy-bodied rodents about 12 inches (30 centimeters) long, not including the long tail. They are native to North America and common to marshes all over the country. Their hind feet are webbed for swimming and they can often be seen floating on the water's surface. Marsh plants such as sedges, reeds, and the roots of water plants provide most of their diet.

Muskrats build dome-shaped houses in water. They pile up mud, cattails, and other plants until the mound rises above the water's surface. Tunnels lead into the mound where one or more rooms are hollowed out above the water's surface.

These mammals are hunted for their fur, which is brown and consists of soft underfur and a dense coat. They are also sold as food, often labeled as "marsh rabbit."

Food Some aquatic mammals, like the otter, are carnivores, eating rabbits, birds, and fish. Muskrats are omnivores, eating both animals, such as mussels, and plants, such as cattails. Beavers are herbivorous and eat trees, weeds, and other plants.

Reproduction Mammals give birth to live young that have developed inside the female's body. Some mammals are helpless at birth while others are able to walk and even run immediately. Some are born with fur and with their eyes and ears open. Others, like the muskrat, are born hairless and blind. After about three weeks young muskrats are able to see and swim.

Endangered species Over half of the endangered or threatened fish and other wildlife in the United States (about 240 species altogether) rely partly on wetlands for food, water, shelter, or a place to reproduce. Over one-third of these live only in wetlands.

In North America, endangered species include the whooping crane and the manatee. The whooping crane lives in coastal swamps and feeds on roots and small reptiles. It has been threatened by hunting, pollution, and dredging (dragging a net along the bottom of a body of water to gather shellfish or plant specimens).

The manatee is a slow-moving, seal-like animal. Manatees live in shallow coastal wetlands in the Caribbean, the Amazon, and Africa. They have become endangered because of overhunting, being caught and strangled in fishing nets, or being killed by boat propellers. Pollution has disrupted their habitats and food sources.

In Ireland, the Greenland white-fronted goose, which relies on bogs for feeding and breeding, is endangered because many bogs have disappeared.

Human Life

After 13000 BC, wetlands played an important role in early civilizations. For prehistoric communities throughout the world they provided food, water, and materials for clothing, shelter, and tools. Wetlands are able to reveal many of these ancient civilizations' secrets to scientists. Bodies of animals and humans, and artifacts (objects made by humans, including tools, weapons, jars, and clothing) have been well preserved in some wetlands and give information about how ancient people lived. Evidence of these ancient communities has been found in bogs. Conditions in bogs are excellent for preserving artifacts and bodies.

Impact of wetlands on human life Wetlands have had an affect on the supply of food and water, on shelter, and on other resources.

The Swimming Rabbit

The North American swamp rabbit has broad, flat feet so that it can move easily over wet, marshy soil. It can swim, and often escapes danger by staying submerged with only the tip of its nose exposed to the air in order to breathe.

Muskrats are residents of marshy areas throughout most of North America. IMAGE COPYRIGHT JUDY CRAWFORD, 2007. USED UNDER LICENSE FROM SHUTTERSTOCK.COM.

The Bog People

Organic materials in peat bogs absorb and hold large amounts of water like a sponge. Bog soils are extremely acidic and have little oxygen. Because of these factors, decomposition takes place very slowly. Entire trees, animals, and even human bodies have been preserved for centuries.

The remains of more than 2,000 humans have been found in bogs, primarily in northwestern Europe. The bodies are in various stages of preservation, from skeletons to those with flesh intact. Sometimes only body parts such as heads or limbs are found. Often the skin has darkened and the hair has turned red from peat acids. The bodies found in Europe range in age from about 1,500 to 3,000 years old. The body of a woman, known as the Koelbjerg Woman, was found in Denmark and is believed to be 10,000 years old.

Many of these people appear to have died violent deaths. Perhaps they were killed as a punishment for a crime or were the victims of human sacrifice. Their deaths may have taken place at the bog because it was so isolated.

At a bog in Windover, Florida, 160 bodies have been found that are about 7,000 years old. Buried with them are food, tools, clothes, and weapons. Many of these items were made from wetland plants and from the bones of wetland animals.

Food and water Wetlands are a source of water for drinking and crops. Areas near wetlands continue to provide food in a variety of ways for tribal cultures in Asia and Africa and for urban peoples elsewhere. Since wetlands are a habitat for much wildlife, hunting and fishing are common there.

Cattle can be raised near wetlands because water and grazing land are available. Crops such as sorghum (a cereal grain native to Africa and Asia) are commonly grown in wetland areas. Grain sorghums, such as milo, kafir, and durra, have adapted to the extremes of the wet/dry cycle of the wetland and are among the most drought-tolerant grains. Millions of people in China, India, and Africa rely on sorghum as a food staple. In the United States, sorghum is used mostly as livestock feed.

Besides sorghum, American farmers use wetland plants, such as marsh grass, reeds, and sedges, for feeding livestock. Wetlands are an excellent place to grow rice, which is a major food source for much of the world's population.

Shelter Materials for dwellings are available from wetland sources. Roofs made from reeds are used on huts in Egypt and in stilt houses in Indonesia to keep the occupants cool and dry. A tightly woven reed roof can last for forty years. House frames are constructed from the timber of mangrove, and palm trees. Wetland sediments, such as clay and mud, may be used to produce bricks that are used for walls.

Other resources   Wetlands provide other resources that humans rely on besides building materials. For example, dried peat is used in homes for heat and to fuel electric generators in countries such as Ireland, where coal is scarce. Peat is a source of protein for livestock feed, and chemically processed peat is used as a base for polishes and waxes. In horticulture, peat moss is used as ground cover, a soil conditioner, and a growing medium. Peat is also found in medicinal baths and cosmetics.

There are at least 120 distinct medicinal substances derived (made) from wetland plants. Wetland soils are sources of gravel and phosphate. Phosphate is used as a raw material for making fertilizers, chemicals, and other commercial products.

Wetlands prevent loss of human life and property during floods because they absorb and store water. The riparian wetlands along the Mississippi River once stored sixty days of floodwater. Since most of these wetlands have been filled or drained, they now only store twelve days of floodwater.

Impact of humans on wetlands   The important role of the wetland for plant and animal communities and for the environment in general has not always been understood. For more than 2,000 years, people in different countries have been draining wetlands to get rid of mosquitoes and disease, to increase land for farming, and to make room for development. By 1990, more than half of the wetlands in the United States were destroyed.

As the value of wetlands becomes more recognized, government agencies, such as the U.S. Department of the Interior and U.S. Fish and Wildlife Service, as well as environmental groups, are working to preserve existing wetlands and to create new ones. In 1989, U.S. president George Bush, asked that the United States work toward the goal of no net loss of wetlands. This means that if a natural wetland is destroyed by development, an artificial one must be built to replace it. As of July 1998, there were 97,000,000 acres (39,000,000 hectares) of artificial wetlands in the United States. In 1998, the Clinton administration issued an

Houses Without Walls

The Seminoles, a Native American people who once lived in the Florida Everglades, built homes with no walls. Called a chikee, the home was built on a platform, had a log framework, and a thatched reed roof. The absence of walls provided ventilation in the warm climate.

An Earth-Friendly Architect

Architect James Cutler of Seattle, Washington, believes that the environment should be preserved when designing and building houses. He has gone to extremes to save the landscape, including building houses on stilts and placing sidewalks over the tops of forests. Cutler designed a home for Microsoft Chairman Bill Gates, where he reestablished wetlands on the property.

Reeds are harvested and bundled to be used on thatched roofs. IMAGE COPYRIGHT MARTIN WALL, 2007. USED UNDER LICENSE FROM SHUTTERSTOCK.COM.

initiative calling for a net gain of 100,000 acres (40,500 hectares) of wetlands per year beginning in 2005.

The Convention on Internationally Important Wetlands was signed by representatives of many nations in Ramsar, Iran, in 1971. Commonly known as the Ramsar Convention, it is an intergovernmental treaty that provides the framework for national action and international cooperation for the conservation and wise use of wetlands and their resources.

Use of plants and animals When animals and plants are overharvested they may become endangered. Overharvesting means they are used up and destroyed at a faster rate than they can reproduce. When this happens, wetland, timber, fuel, medicines, and sources of food for humans and animals are all lost.

Use of natural resources The two primary resources found in wetlands are water and peat. Wetlands often act as part of the groundwater system. When they are lost, the supply of drinking water may be affected.

It takes ten years to form less than half an inch of peat, and its overuse has caused significant peatland losses. Western European peat mining companies are rapidly using up local areas of peat and expanding their mining operations into eastern European countries. All natural peatlands in the Netherlands have been destroyed. Switzerland and Germany each have only 1,250 acres (500 hectares) of peatland remaining.

Quality of the environment Wetlands are endangered by industrial and municipal (city) contaminants, by the accumulation of toxic chemicals, and from acid rain.

Acid rain is a type of air pollution especially dangerous to wetlands. It forms when industrial pollutants such as sulfur or nitrogen combine with moisture in the atmosphere and form sulfuric or nitric acids. These acids can be carried long distances by the wind before they fall either as dry deposits or in the form of rain or snow. Acid rain can significantly damage both plant and animal life. It is especially devastating to wetland amphibians such as salamanders, because it prevents their eggs from maturing.

Mining for minerals near a wetland can have a negative impact on the biome. Mercury, a poisonous liquid metal used in gold mining operations, often contaminates wetlands close to the mines. Mining operations also require large amounts of water, and nearby wetlands may be drained.

The world's climate may be growing warmer because of human activity. As polar ice caps melt, there is a rise in sea level. As this happens, more salt water floods into coastal wetlands and increases the salinity not only of the wetlands, but of rivers, bays, and water supplies beneath the ground. As these habitats change, animal and plant life are affected, and some wetlands may be destroyed.

Artificial wetlands Artificial wetlands are those created by humans. Creating a wetland is very difficult; all the right conditions need to be met, including a water source and nutrient rich soil that contains a lot of water but not much oxygen. In Arizona, there are twenty-six artificial wetlands operational with twenty-four others under construction or awaiting approval.

An artificial wetland that works well and is extremely valuable worldwide is a rice paddy. A rice paddy is simply a field that is flooded for the purpose of growing rice, a food staple for about three billion people—nearly one half of the world's population. Some are flooded naturally, by monsoon (tropical) rains or overflowing rivers. Others are flooded by irrigation (watering). Mud dams and waterwheels are built to bring in and hold the water level at approximately 4 to 6 inches (10 to 15 centimeters) while the rice grows. Ninety percent of rice paddies are in Asia, especially China and India. Some rice is also grown in Europe and the United States.

Artificial wetlands are found on the Arabian Peninsula where they are used for water storage and sewage treatment. Other human-made wetlands have been created as winter homes for migrating waterfowl.

Adopt A Wetland

Both the federal government and individual states are working to protect wetlands. Some states sponsor "Adopt a Wetland" programs, in which groups of people agree to help support their local wetlands in a variety of ways, such as picking up litter. For information about adopting a wetland, contact the U. S. Environmental Protection Agency, Public Information Center, (202) 260-7756 or (202) 260-2080.

Marshes Clean Up Messes

A lot of contaminants seep into the environment every day from oil spills, sewage treatment plants, industrial waste, and old mines. In some areas, phytoremediation (fi-toh-rih-mee-dee-AY-shun) is being used to clean up these messes. (*Phyto* is a Greek word for plant and *remediation* means to make things right.) Phytoremediation uses plants to absorb pollutants.

One type of phytoremediation involves creating an artificial marsh containing cattails and water lilies. These wetland plants are able to absorb some water pollutants. Poplar trees can clean up water polluted with oil, and sunflowers are used to help clean up radioactive materials in the soil. Sunflowers are used near the Chernobyl nuclear power station in Pryp'yat near Kiev in the Ukraine. The power station exploded in 1986, releasing radioactive waste into the environment.

Many varieties of rice grow well in shallow water. IMAGE COPYRIGHT JAPONKA, 2007. USED UNDER LICENSE FROM SHUTTERSTOCK.COM.

Native peoples   The Marsh Arabs of Iraq and the Nilotic peoples of Africa live in or near wetlands that provide almost all of the resources they need.

Marsh Arabs of Iraq   The Mesopotamian wetland lies between the Tigris and Euphrates Rivers in southern Iraq. At almost 8,000 square miles (20,000 square kilometers), it is one of the largest wetlands in the world. The Marsh Arabs, or Ma'dan, have lived there for more than 6,000 years. The water there is clean, calm, and fairly shallow—about 8 feet (2.4 meters) deep.

The Marsh Arabs' houses are made of reeds and built on islands that they make. To construct the islands, they create a fence from reeds and partially submerge it. Then they fill the area inside the fence with cut rushes, add layers of mud, and stamp it all down. When the pile reaches the water's surface, they fold the top of the fence onto the pile and add more reeds to finish it. The completed island is big enough not only for the family to build a house and live on, but for their cattle as well.

Isolation from the outside world has allowed the Marsh Arabs' way of life to remain the same for hundreds of years. Daily life consists of fishing, buffalo herding, and growing rice. Reeds are gathered every day to feed the buffalo. Transportation from house to house or to other villages is by means of small canoes called *mashhufs*.

Survival of the Marsh Arabs is threatened by irrigation practices, which have drawn water from their wetlands. Some of the marshes have already been drained for agricultural use and oil exploration.

Nilotic Peoples   Nilotic peoples live near the Nile River or in the Nile Valley in Africa. Two Nilotic tribes, the Dinka and the Nuer, live in southern Sudan. This area is a rich floodplain, and during the wet season, from July through October, the people live on high ground in permanent villages. From December through April, when the floodwater recedes, they move to the floodplains. There, the Dinka build temporary villages along the banks of the Nile River. Their homes, made from reeds and grasses, are circular at the base and come to a point at the top, much like a teepee. Their neighboring tribe, the Nuer, lives in a similar fashion but in the marsh, savanna (grassland), and swamp areas.

Madan, or Marsh Arab, children play in water canals that still are left in what is now mostly dry land in Qurnah, southern Iraq. AP IMAGES.

Wetland grasses and plants grow in the floodplain after the waters recede. The grasses provide food for cattle and attract wildlife, which is often hunted for food. Nilotic tribes fish, hunt, and grow some grains. The Dinka use their cattle for meat and milk, but the Nuer eat cattle only during religious ceremonies. For them, cattle represent wealth and are used for bridal dowries (gifts) and to sell. Animal hides are made into clothing and bedding, and dried cattle dung (waste) is used for fuel.

The Food Web

The transfer of energy from organism to organism forms a series called a food chain. All the possible feeding relationships that exist in a biome make up its food web. In the wetland, as elsewhere, the food web consists of producers, consumers, and decomposers. These three types of organisms transfer energy within the wetland environment.

At the bottom of the food chain are the producers, such as photosynthesizing plants and algae, which use energy from the sun to produce sugars and starch from carbon dioxide and water. The shallow wetland water lets in lots of sunlight, which helps these organisms grow.

Primary producers are eaten by primary consumers such as the larval forms of frogs and toads, and larger animals such as shrimp and snails. Other primary consumers, such as small aquatic insects, shellfish, and small fish, feed on plant materials.

Primary consumers are food for predators such as larger fish, reptiles, amphibians, birds, and mammals, which are called secondary consumers. Tertiary consumers are the predators, such as owls, coyotes, and humans, that prey on both primary and secondary consumers.

The decomposers feed on dead organic matter and include fungi, bacteria, and crabs.

Spotlight on Wetlands

The Florida Everglades   The Florida Everglades, lying within the subtropical zone near the tropics, is one of the largest freshwater marshes in the world, stretching from Lake Okeechobee 100 miles (160 kilometers) south to Florida Bay. Within its boundaries are freshwater swamps and coastal (saltwater) mangrove swamps. In the northern portion of the Everglades is Big Cypress Swamp. This swamp covers 1,500 square miles (4,000 square kilometers) and gets its name from the many tall bald cypress trees that live there.

The Everglades were formed during the last ice age, which ended about 10,000 years ago. During this period, glaciers melted and sea level was raised, which flooded the area and turned it into a wetland. Under the water and soil is a porous limestone rock formed during the last glacial period. This type of rock contains shells and skeletons of animals deposited in the sea. Hummocks (rounded hills or ridges) have formed in some areas, and hardwood trees can grow there. Otherwise, the terrain is flat.

The subtropical climate of the Everglades is mild, with temperatures ranging from 73° to 95°F (25° to 35°C) in the summer. Winters are mild, with an average high of 77°F (25°C) and a low of 53°F (12°C). The rains occur from June to October, with an annual accumulation of about 55 inches (140 centimeters). The region is often hit by severe tropical storms.

The grassy waters of the Everglades are actually very shallow, ranging from about 6 inches (15 centimeters) to 3 feet (94 centimeters) deep. They cover about 4,000 square miles (10,300 square kilometers). Water comes from rainfall and overflows from Lake Okeechobee.

Much of the Everglades are covered with sawgrass, a type of sedge. Tropical plants, such as ferns, orchids, and mosses, grow in freshwater areas. Saltwater aquatic plants include lilies and bladderworts. Cypresses, palms, live oaks, and pines grow on the hummocks.

Among the invertebrates, the Liguus tree snail can be found living on the hummocks, and Everglades reptiles include turtles, king snakes, water moccasins, and rattlesnakes. Alligators live in freshwater areas while American crocodiles live saltwater swamps.

Tourists are attracted by the bird life in the Everglades. Herons, egrets, spoonbills, ibises, eagles, and kites are a few of the birds in the area. The Everglades kite, a tropical bird of prey, has been named after the region. Mammals include white tailed deer, cougars, bobcats, black bears, and otters.

The Florida Everglades
Location: Southern Florida
Area: Approximately 1,506,539 acres (606,688 hectares)
Classification: Freshwater marsh, freshwater swamp, and saltwater swamp

Many attempts have been made to drain the Everglades so the land could be used for farming and development. Drainage projects began in the early 1900s. Consequently, about 50 percent of the wetlands have been altered. Due to agriculture and development, only about 27 percent of the original area has been preserved as the Everglades National Park.

Draining the area has caused a loss of animal and plant life. Endangered species include the manatee, the round-tailed muskrat, and the mangrove fox squirrel. Early in the twentieth century, plume hunters nearly wiped out egrets and spoonbills for their feathers, which were used extensively as decorations on women's hats.

The introduction of non-native species of plants is also a threat. The Australian melaleuca tree and the Brazilian pepper plant have spread throughout the area, using up water supplies and choking native plants.

Agriculture is the leading economic activity in the region. In the area south of Lake Okeechobee, farmers raise sugarcane, fruit, and vegetables, which are shipped north. Sport fishing, boating, hunting, and camping are popular in the area.

The Seminole Indians, who were driven out of the Okefenokee Swamp in northern Florida by American troops during the nineteenth century, found a safe home in the Everglades. They planted corn and vegetables and gathered roots and nuts. Fishing and hunting were also sources of food. Throughout the twentieth century, the number of Seminoles living in the area has decreased. Another tribe, the Miccosukee, still live in reservations near the center of the Everglades.

A growing concern for the environment has encouraged protection of the Everglades and its ecosystem. It became a national park in 1947 and is the third largest and the wettest national park in the United States. The park includes 10,000 islands off the Florida coast along the Gulf of Mexico, and more than one million people visit annually.

Okefenokee Swamp   The word "Okefenokee" (oh-kee-fen-OH-kee) is taken from a Timucuan Indian word and means "trembling earth." This refers to the small bushes and water weeds that float on the swamp's open water and move when the water is disturbed. The swamp, which is 25 miles (40 kilometers) wide and 38 miles (61 kilometers) long, is one of the oldest and most well-preserved freshwater wetlands in the United States.

The Okefenokee lies about 50 miles (80 kilometers) inland from the coast of the Atlantic Ocean in Georgia, but a sandy ridge prevents the

Okefenokee Swamp
Location: Southeastern Georgia and northern Florida
Area: 438,000 acres (177,000 hectares)
Classification: Freshwater swamp, bog, marsh

swamp from draining directly into the ocean. The Suwannee River is the principal outlet of the swamp providing 85 percent drainage into the Gulf of Mexico.

The Okefenokee was formed almost 250,000 years ago when the waters of the Atlantic Ocean covered the area. When the ocean receded, some salt water was trapped in depressions in the ground. Over thousands of years, plants grew and filled the wetland. As they decayed, they turned into peat, which now forms the swamp floor. The peat is as deep as 15 feet (4.6 meters) in places.

Freshwater lakes, wet savannas (grasslands with scattered trees), cypress woods, hummocks, open grassy spaces, peat bog islands, and thick brush are also found in the Okefenokee. Many channels form a maze through the area. About 60 inches (152 centimeters) of rain falls annually.

The Okefenokee supports 621 species of plants, including rare orchids and lilies. The insectivorous pitcher plant, maiden cane, floating hearts, and golden club all add color. Freshwater regions support sphagnum moss, ferns, and rushes. Tupelo trees and giant bald cypress trees covered with Spanish moss add an eerie quality.

Insects, such as the mosquito and the yellow fly, are abundant, and 37 species of amphibians make the Okefenokee their home. At least 64 species of reptiles live here, including 5 kinds of poisonous snakes, and more than 10,000 alligators. Considered "kings" of the swamp, alligators are a major tourist attraction. About 40 species of fish live in the swamp, including bluegill, warmouth, golden shiners, and pumpkinseed sunfish. These fish have adapted to life in the acidic water. They are often dark with yellow undersides and they can see in low light.

The Okefenokee is home to 233 species of birds including the white ibis, the sandhill crane, the wood duck, the great blue heron, and the barred owl.

Bears, bobcats, deer, foxes, and raccoons live in the forested areas. Approximately thirty species of mammals can be found in the swamp, including the black bear.

The area surrounding the swamp was settled by Europeans in the late 1700s. The Seminoles once lived in the interior, but they were driven out by American troops in 1838.

During the 1800s, developers tried to drain the swamp by building a canal, but found the project larger than anticipated. The project was abandoned in 1893. After the turn of the century, logging companies built railroad tracks through the swamp; as a result 400,000,000 feet (120,000,000 meters) of timber, mostly cypress were cut down. In 1937, U.S. president Franklin Roosevelt (1882–1945) protected about 629 square miles (1,630 square kilometers) of the Okefenokee by designating it the Okefenokee National Wildlife Refuge. Despite these actions, the swamp continues to be threatened by developers, including those who want to mine titanium ore on its eastern rim. Due to public and government opposition, the mining project was terminated.

Fire is a natural part of the swamp's ecosystem, which can occur as infrequently as every 30-50 years. The two most recent fires occurred in May 2002 and May 2007.

The Okefenokee offers walking trails for tourists and 120 miles (193 kilometers) of water trails that can be explored by canoe.

The Mekong Delta   The Mekong Delta extends from Phnom Penh, the capital of Kampuchea province in Vietnam, to the coastline bordering the South China Sea. The Mekong River, the longest in Southeast Asia, divides at Phnom Penh into two branches. As the branches pass through the delta they form nine channels. The name Mekong means "Nine Dragons" and refers to these channels.

The delta, which supports 15,000,000 Vietnamese, forms a 13,600,000 acre (5,500,000 hectares) triangular area including floodplain, freshwater, and saltwater swamps. Mangrove and melaleuca forests cover the remaining area.

Most of the delta is less than 16 feet (5 meters) above sea level. From May to October the area experiences heavy rains, ranging from 59 to 92 inches (150 to 235 centimeters). The average temperature is about 79°F (26°C), and the climate is very humid.

The Mekong Delta supports 35 species of reptiles, including crocodiles and the endangered river terrapin (a type of turtle), and 260 species of fish, many of which are sold commercially. Herons, egrets, ibises, and storks nest in the delta's forested areas. Mammals include otters and fishing cats.

The delta is rich in agricultural produce, fish, and waterfowl. At one time it yielded 26 tons (23.5 metric tons) of fish per square mile. About half of the delta is devoted to rice paddies, which yield about 6,500,000 tons (5,900,000 metric tons) of rice each year.

During the Vietnam War (1959–1975), the Mekong Delta was seriously damaged when Agent Orange and other defoliants (chemicals

The Mekong Delta
Location: Vietnam
Area: 12,237 acres
(49,712 square kilometers)
Classification: Freshwater swamp and saltwater swamp

that cause leaves to fall off plants and trees) destroyed over 50 percent of the mangrove forests. The object was to remove any hiding places used by the enemy, but Agent Orange also caused skin diseases, cancer, and birth defects. One-fifth of the country's farmland was destroyed by direct bombing and machines used to clear land. The melaleuca forests were burned with Napalm, another chemical weapon, and almost destroyed. These chemicals made much of the soil infertile.

Of the 386 bird species supported by the Mekong Delta, 92 are waterfowl. The eastern Sarus crane, an endangered bird, disappeared during the war but has since returned. The giant ibis, the white-shouldered ibis, and the white-winged wood duck, all endangered, have not been seen here since 1980.

Hydroelectric dams, built for generating electricity, have a negative affect on the ecosystem. The dams reduce flooding, which decreases the freshwater flow to wetlands. Spawning areas for fish are declining and mangrove forests are being converted to artificial ponds by the farmed shrimp industry.

The Great Dismal Swamp   The Great Dismal Swamp is a freshwater wetland located in southeast Virginia and northeast North Carolina. Its soil contains few minerals and nutrients. Drainage is poor and the soil is saturated and acidic. In this environment, decomposition is slow and peat accumulates over time.

The fungus called armillaria, which grows on rotten wood in the swamp, gives off a luminescence (glow) called foxfire. Broad-leaved shrubs such as hollies and bayberries are found here. Bald cypress trees survive well in saturated soil and populate the area; some are as old as 1,500 years. Other native trees include the black gum, the juniper, and the white ash.

Seventy-three species of butterflies and more than 97 species of perching birds, including warblers, woodpeckers, and wood ducks, live in this area. Black bears, bobcats, minks, and otters also live in the region.

Many famous Americans were in some way associated with the Great Dismal Swamp. In 1763, U.S. president George Washington (1732–1799) set up a company to drain the swamp for agricultural purposes and to use its lumber. One hundred-forty-five miles of road were constructed to support the logging activities that continued until 1976. The American statesman, Patrick Henry (1736–1799), who participated in the American

The Great Dismal Swamp
Location: North Carolina and Virginia
Area: 106,716 acres (42,686 hectares)
Classification: Peat bog and freshwater swamp

Revolution, owned land there. The swamp was used by American author Harriet Beecher Stowe (1811–1892) as the setting for her 1856 novel, *Dred: A Tale of the Great Dismal Swamp.* The forests of the swamp provided refuge to runaway slaves. In 2003, the Great Dismal Swamp became the first National Wildlife Refuge to be officially designated as a link in the Underground Railroad Network to Freedom.

Kakadu National Park   Kakadu National Park is the largest wetland in Australia and is marked by a great diversity in species and habitats. The area includes freshwater wetlands, mangrove swamps, and billabongs. The term billabong comes from an Australian Aboriginal (native) word meaning "dead river," and describes standing, often stagnant, water near river channels. Billabongs are usually found only in the parts of Australia where the climate is hot and dry and interrupted by occasional flooding.

Average temperatures are high all year and range from 91°F (33°C) in July to 108°F (42°C) in October. Rivers flooded by January and February rains feed the freshwater areas. The coastal wetlands are fed by tides.

Kakadu includes many acres of forests, sedges, and grasslands. In April and May, when the dry season begins, bush fires destroy old growth and encourage new plant life.

When the thousands of acres of grassland flood during the wet season, plant and animal life abounds. In all, there are more than 200 species of plants, including 22 species of mangrove trees. Water lilies, woolybutt (*Eucalyptus miniata*), and spear grass, which grows to more than 6 feet (1.8 meters) tall, can all be found. Paperbark trees dominate freshwater swamp areas.

Numerous turtles and both freshwater and saltwater crocodiles live in the shallows. Barramundi fish travel between the waterholes and estuaries, depending on the season. Goannas, a type of monitor lizard, live here.

Millions of water birds live in the park. These include whistling ducks, magpie geese, radjal shelducks, and grey teals. Shorebirds migrate from as far away as the Arctic to winter in the mild climate of Kakadu.

The Australian native people, called Aborigines, once used the park's wetland plants and animals as a source of food. Aborigines recognized six seasons in the Kakadu region instead of the traditional wet and dry seasons. Disposal of waste from uranium mines on the edge of Kakadu is threatening the area. Other dangers include tourism and the introduction of non-native plants and animals. Para grass, an exotic grass being

Kakadu National Park
Location: East of Darwin in the Northern Territory of Australia
Area: 7,700 square miles (20,000 square kilometers)
Classification: Freshwater wetlands, mangrove swamps

melted. The many lakes in the area are in the process of turning into bogs. Sandy lowlands can be found among a maze of rivers, and pine forests and floodplains add to the diversity of the ecosystem. About one-third of the region is forested.

The climate is cool, with winter temperatures ranging from 18° to 25°F (–8° to –4°C). Warmer temperatures are found in and around the marshes. Annual precipitation is 22 to 26 inches (55 to 65 centimeters).

Trees growing here include pine, birch, alder, oak, aspen, white spruce, and hornbeam. Hornbeam is a type of birch that has smooth, gray bark and catkins (drooping scaly flower clusters with no petals).

Many types of birds live in the marsh, including orioles, grouse, woodpeckers, owls, blue tits, and ducks. Mammals include lynxes, wolves, foxes, wild boars, Asian deer, beavers, badgers, and weasels.

Much of the land has been cleared for lumber and agricultural use over the last several hundred years. Large-scale land reclamation (draining swampland to create cropland) in the twentieth century occurred to promote the development of agricultural areas. Crops grown in the area include rye, barley, wheat, flax, potatoes, and vegetables. Grasses used as cattle feed are also grown here.

The Pine Barrens   The Pine Barrens (or Pineland) are located on the outer coastal plain of Long Island and New Jersey, including parts of the Delaware River and the Jersey Shore. Much of the area is open forest broken by marshes, swamps, and bogs. The area was formed during the last Ice Age.

The northwest part of the Pine Barrens experiences relatively cold winters, with average January temperatures of less than 28°F (–2°C). The southern area is milder, with average winter temperatures above freezing. Summers are hot, with averages for July ranging from about 70°F (21°C) in the northwest to more than 76°F (24°C) in the southwest. Precipitation (rain, sleet, or snow) is evenly distributed throughout the seasons, averaging between 44 and 48 inches (112 and 122 centimeters) annually.

The Pine Barrens are home to about 100 endangered species of plants. Many rare species include the glade cress, great plains ladies-tresses, grooved yellow flax, and some insectivorous plants. Other plants include wild azaleas, purple cone flowers, Indian grasses, and little blue-stems. Rhododendrons, honeysuckles, mountain laurels, wintergreens, and cardinal flowers bring color to the area.

The Pine Barrens
Location: Long Island, New Jersey
Area: Over 1,000,000 acres (400,000 hectares)
Classification: Bog, swamp, and marsh

The area is dominated by oak and pine trees, which thrive on the well-drained sites. White cedars grow in the poorly drained bogs. Other trees include buckthorns, dogwoods, sugar maples, hemlocks, birches, ashes, and sweet gums.

The endangered Pine Barrens tree frog makes its home in the park, while bears and wildcats can still be found in some of the woodland areas. Deer, opossums, and raccoons are common.

Commercially, the area is valued for its production of blueberries and cranberries, and for tourism.

Designated by Congress in 1978 as the country's first National Reserve, the Pine Barrens' natural and cultural resources are now protected.

On May 1, 2007, a flare from an F-16 plane accidentally dropped into the forest during training, resulting in a fire that burned more than 17,000 acres.

Polar Bear Provincial Park Polar Bear Provincial Park lies along the southern edge of the Arctic region, on the northwest coast of James Bay and the southern coast of Hudson Bay in Ontario, Canada. A true wilderness, it is accessible only by plane or boat. The dominant type of wetland in the park is peatland. Inland areas of the park contain swamps and marshes, while coastal areas contain saltwater marshes.

Many of the wetlands have formed in kettles (depressions in the ground left behind by retreating glaciers). Rain, melted snow, and over-flowing rivers contribute to the water supply. As kettle lakes fill in with vegetation, marshes are formed, and coastal marshes have developed with the movement of the tides. The flat-topped ridges that run along the beach, parallel to the coast, help prevent drainage back into the sea.

Because the park lies along the southern edge of the Arctic region, temperatures are cold. Freezing weather prevails for six to eight months. Summers are short with the average temperature between 60° and 70°F (16° and 25°C). Precipitation is usually less than 21 inches (55 centimeters).

Vegetation varies depending on altitude and wetness. In the higher, cooler regions of the park, plant life includes sedges, goose grasses, and flowering saxifrages. Cotton grasses, sedges, and birches grow in the lower areas. In drier places, blueberries, crowberries, and louseworts are found. Salt-tolerant plants grow in the saltwater marshes, including aquatic grasses, cotton grasses, lungworts, and lyme grasses. Caribou and reindeer lichens grow in wetlands closer to forested areas.

Polar Bear Provincial Park
Location: Ontario, Canada
Area: 9,300 square miles (24,087 square kilometers)
Classification: Peatland, swamp, freshwater marsh, and saltwater marsh

Many waterfowl, such as the lesser snow goose, use the park in spring and fall as they migrate to their Arctic breeding grounds. Thousands of mallard ducks pass through in the fall, and the western and southwestern parts of the park form a migration path for shorebirds, such as the ruddy turnstone, black-bellied plover, and several species of sandpipers. Canada geese nest in the park.

Polar bears spend their summers in the park while breeding. Caribou and moose roam the area, moving north in the warmer weather. Beavers, muskrats, otters, foxes, and wolves are just a few of the many other mammals that make the park their year-round home.

The Cree Indians live along the coastal areas and use the park for hunting, fishing, and trapping. The Cree own two hunting and fishing camps where guests can fish and hunt waterfowl, grouse, and snipe. No other non-native hunting or fishing is permitted.

Since 1970 the area has been protected from development. Commercial or industrial use of natural resources is prohibited. Most of the park is designated as wilderness zones, nature reserves, or historical zones, where wildlife is protected.

Usumacinta Delta   The Usumacinta (oo-soo-mah-SEEN-tah) Delta is the most extensive wetland on the Gulf Coast of Mexico. It contains freshwater lagoons, swamps, marshes, and mangrove swamps. Its waters are rich in nutrients and support a major coastal fishery.

Many waterbirds breed and winter in the area. These include herons, egrets, storks, ibises, and spoonbills. Shorebirds that winter in the delta are ducks and coots.

A few manatees from the West Indies can be found here, as well as the endangered Morelet's crocodile.

Threats to the delta have come from drainage for agriculture and oil spills from a nearby oil field. Also, mangrove trees are being cut for timber.

Flow Country   The marshy area of Caithness and Sutherland countries is known as Flow Country. The name "flow" comes from an old Norse word meaning "marshy ground." Flow Country is one of the largest blanket bogs in the world, and it is still growing. Some of the peat found here is more than 8,000 years old.

Typical plants that grow in Flow Country include sphagnum moss, heather, purple moorgrass, sedges, and rushes. The insect-eating plant, sundew, inhabits the bog.

Usumacinta Delta
Location: Gulf Coast of Mexico
Area: 2,470,000 acres (998,000 hectares)
Classification: Swamp, marsh, and mangrove swamp

Flow Country
Location: Northeast corner of Scotland
Area: 902,681 square acres (365,310 hectares)
Classification: Blanket bog

Many bird species are supported by the bog. Sixty-six percent of the European communities' greenshanks and the entire population of black-throated divers live there.

The peat that grows in a blanket bog is very valuable as fuel for home heating and industrial use. Peat mining is endangering this ancient ecosystem. In the 1980s, the Scottish government and several private companies began planting pine and fir trees. The trees support the forestry industry, but will eventually take over the wetland. Planting was halted by 1990 and the land was purchased by the Royal Society for the Protection of Birds (RSPB) in 1995 to preserve 14,800 acres (6,000 hectares) of forest.

For More Information

BOOKS

Batzer, Darold P., and Rebecca R. Sharitz. *Ecology of Freshwater and Estuarine Wetlands.* Berkeley: University of California Press, 2007.

Cox, Donald D., and Shirley A. Peron. *A Naturalist's Guide to Wetland Plants: An Ecology for Eastern North America.* Syracuse, NY: Syracuse University Press, 2002.

Fowler, Theda Braddock. *Wetlands: an Introduction to Ecology, the Law, and Permitting.* Lanham, MD: Government Institutes, 2007.

Mitsch, William J., and James G. Gosselink. *Wetlands.* 4th ed. Hoboken, NJ: Wiley: 2007.

Moore, Peter D. *Wetlands.* New York: Chelsea House, 2006.

Scrace, Carolyn. *Life in the Wetlands.* New York: Children's Press, 2005.

Shea, John F., et al. *Wetlands, Buffer Zones and Riverfront Areas: Wildlife Habitat and Endangered Species.* Boston: MCLE, 2004.

Swarts, Frederick A. *The Pantanal: Understanding and Preserving the World's Largest Wetland.* St. Paul, MN: Paragon House Publishers, 2000.

PERIODICALS

Johnson, Dan. "Wetlands: Going, Goings ... Gone?" *The Futurist.* 35. 5 September 2001: 6.

Levathes, Louise E. "Mysteries of the Bog." *National Geographic.* Vol. 171, No. 3, March 1987, pp. 397–420.

Mohlenbrock, Robert H. "Wetted Bliss: in a Louisiana Refuge, Different Degrees of Moisture Create Distinctive Woods." *Natural History.* 117. 2 March 2008: 62.

Yalden, Derek. "Managing Moorland." *Biological Sciences Review.* 18. 2 November 2005: 14.

ORGANIZATIONS

Environmental Protection Agency, Office of Wetlands, Oceans and Watersheds, Wetlands Division (4502F), 401 M Street SW, Washington, DC 20460, Phone: 202-260-2090, Internet: http://www.epa.gov.

National Wetlands Conservation Project, The Nature Conservancy, 1800 N Kent Street, Suite 800, Arlington, VA 22209, Phone: 800-628-6860; Internet: http://www.tnc.org.

WEB SITES

"America's Wetlands." Environmental Protection Agency. http://www.epa.gov/OWOW/wetlands/vital/what.html (accessed July 13, 2007).

"On Peatlands and Peat." International Peat Society. http://www.peatsociety.fi (accessed July 13, 2007).

"Status and Trends of Wetlands of the Counterminous United States from 1998 to 2004." U.S. Fish & Wildlife Service. http://wetlandsfws.er.usgs.gov/status_trends/National_Reports/trends_2005_report.pdf (accessed July 13, 2007).

Thurston High School, Biomes: http://ths.sps.lane.edu/biomes/index1.html (accessed July 13, 2007).

University of California at Berkeley: http://www.ucmp.berkeley.edu/glossary/gloss5/biome/index.html (accessed July 13, 2007).

Where to Learn More

Books

Allaby, Michael. *Biomes of the Earth: Deserts.* New York: Chelsea House, 2006.

Allaby, Michael. *Biomes of the Earth: Grasslands.* New York: Chelsea House, 2006.

Allaby, Michael. *Biomes of the Earth: Temperate Forests.* New York: Chelsea House, 2006.

Allaby, Michael. *Biomes of the Earth: Tropical Forests.* New York: Chelsea House, 2006.

Allaby, Michael. *Temperate Forests.* New York: Facts on File, 2007.

Angelier, Eugene, and James Munnick. *Ecology of Streams and Rivers.* Enfield, New Hampshire: Science Publishers, 2003.

Ballesta, Laurent. *Planet Ocean: Voyage to the Heart of the Marine Realm.* National Geographic, 2007.

Batzer, Darold P., and Rebecca R. Sharitz. *Ecology of Freshwater and Estuarine Wetlands.* Berkeley: University of California Press, 2007.

Braun, E. Lucy. *Deciduous Forests of Eastern North America.* Caldwell, NJ: Blackburn Press, 2001.

Brockman, C. Frank. *Trees of North America: A Guide to Field Identification,* Revised and Updated. New York: St. Martin's Press, 2001.

Burie, David, and Don E. Wilson, eds. *Animal.* New York: Smithsonian Institute, 2001.

Carson, Rachel L. *The Sea Around Us*, Rev. ed. New York: Chelsea House, 2006.

Cox, C.B., and P.D. Moore. *Biogeography and Ecological and Evolutionary Approach*, 7th ed. Oxford: Blackwell Publishing, 2005.

Cox, Donald D. *A Naturalist's Guide to Seashore Plants: An Ecology for Eastern North America.* Syracuse, NY: Syracuse University Press, 2003.

Day, Trevor. *Biomes of the Earth: Lakes and Rivers.* New York: Chelsea House, 2006.

Day, Trevor. *Biomes of the Earth: Taiga.* New York: Chelsea House, 2006.

Fisher, William H. *Rain Forest Exchanges: Industry and Community on an Amazonian Frontier.* Washington DC: Smithsonian Institution Press, 2000.

Fowler, Theda Braddock. *Wetlands: an Introduction to Ecology, the Law, and Permitting.* Lanham, MD: Government Institutes, 2007.

Gaston, K.J., and J.I. Spicer. *Biodiversity: An Introduction*, 2nd ed. Oxford: Blackwell Publishing, 2004.

Gleick, Peter H., et al. *The World's Water 2004-2005: The Biennial Report on Freshwater Sources.* Washington DC: Island Press, 2004.

Gloss, Gerry, Barbara Downes, and Andrew Boulton. *Freshwater Ecology: A Scientific Introduction.* Malden, MA: Blackwell Publishing, 2004.

Grzimek, Bernhard. *Grizmek's Animal Encyclopedia*, 2nd ed. Vol 7. *Reptiles,* edited by Michael Hutchins, James B. Murphy, and Neil Schlager. Farmington Hills, MI: Gale Group, 2003.

Hancock, Paul L., and Brian J. Skinner, eds. *The Oxford Companion to the Earth.* New York: Oxford University Press, 2000.

Harris, Vernon. *Sessile Animals of the Sea Shore.* New York: Chapman and Hall, 2007.

Hauer, Richard, and Gary A. Lamberti. *Methods in Stream Ecology*, 2nd ed. San Diego: Academic Press/Elsevier, 2007.

Hodgson, Wendy C. *Food Plants of the Sonoran Desert.* Tucson: University of Arizona Press, 2001.

Houghton, J. *Global Warming: The Complete Briefing.* 3rd ed. Cambridge: Cambridge University Press, 2004.

Humphreys, L.R. *The Evolving Science of Grassland Improvement.* Cambridge, UK: Cambridge University Press, 2007.

Hurtig, Jennifer. Deciduous Forests. New York: Weigl Publishers, 2006.

Irish, Mary. *Gardening in the Desert: A Guide to Plant Selection & Care.* Tucson: University of Arizona Press, 2000.

Jacke, Dave. *Edible Forest Gardens.* White River Junction, VT: Chelsea Green Publishing Co., 2005.

Johansson, Philip. The Temperate Forest: A Web of Life. Berkely Heights, NJ: Enslow Publishers, 2004.

Johnson, Mark. *The Ultimate Desert Handbook: A Manual for Desert Hikers, Campers and Travelers.* Camden, Maine: Ragged Mountain Press/McGraw Hill, 2003.

Jose, Shibu, Eric J. Jokela, and Deborah L. Miller *The Longleaf Pine Ecosystem: Ecology, Silviculture, and Restoration.* New York: Springer, 2006.

Luhr, James F., ed. *Earth.* New York: Dorling Kindersley in association with The Smithsonian Institute, 2003.

Marent, Thomas, and Ben Morgan. *Rainforest.* New York: DK Publishing, 2006.

McGonigal, D., and L. Woodworth. *Antarctica: The Complete Story.* London: Frances Lincoln, 2003.

Miller, James H., and Karl V. Miller. *Forest Plants Of The Southeast And Their Wildlife Uses.* Athens: University of Georgia Press, 2005.

Mitsch, William J., and James G. Gosselink. *Wetlands,* 4th ed. Hoboken, NJ: Wiley: 2007.

Moore, Peter D. *Biomes of the Earth: Tundra.* New York: Chelsea House, 2006.

Moore, Peter D. *Wetlands.* New York: Chelsea House, 2006.

Moul, Francis. *The National Grasslands: A Guide to America's Undiscovered Treasures.* Lincoln: University of Nebraska Press, 2006.

National Oceanic and Atmospheric Administration *Hidden Depths: Atlas of the Oceans.* New York: HarperCollins, 2007.

Ocean. New York: DK Publishing, 2006.

O'Shea, Mark, and Tim Halliday. *Smithsonian Handbooks: Reptiles and Amphibians.* New York: Dorling Kindersley, 2001.

Oliver, John E., and John J. Hidore. *Climatology: An Atmosperic Science,* 2nd ed. Upper Saddle River, NJ: Prentice Hall, 2004.

Peschak, Thomas P. *Wild Seas, Secret Shores of Africa.* Cape Town: Struik Publishers, 2008.

Postel, Sandra, and Brian Richter *Rivers for Life: Managing Water For People And Nature.* Washington, DC: Island Press, 2003.

Preston-Mafham, Ken, and Rod Preston-Mafham. *Seashore.* New York: HarperCollins Publishers, 2004.

Primack, Richard B., and Richard Corlett. *Tropical Rain Forests: An Ecological and Biogeographical Comparison.* Hoboken, NJ: Wiley-Blackwell, 2005.

Romashko, Sandra D. *Birds of the Water, Sea, and Shore.* Lakeville, MN: Windward Publishing, 2001.

Scientific American, ed. *Oceans: A Scientific American Reader.* Chicago: University Of Chicago Press: 2007.

Scrace, Carolyn. *Life in the Wetlands.* New York: Children's Press, 2005.

Shea, John F., et al. *Wetlands, Buffer Zones and Riverfront Areas: Wildlife Habitat and Endangered Species.* Boston: MCLE, 2004.

Sowell, John B. *Desert Ecology.* Salt Lake City: University of Utah Press, 2001.

Sverdrup Keith A., and Virginia Armbrust. *An Introduction to the World's Oceans.* Boston: McGraw-Hill Science, 2006.

Swarts, Frederick A. *The Pantanal: Understanding and Preserving the World's Largest Wetland.* St. Paul, MN: Paragon House Publishers, 2000.

Vandermeer, John H., and Ivette Perfecto *Breakfast Of Biodiversity: The Political Ecology of Rain Forest Destruction.* Oakland, CA: Food First Books, 2005.

Vogel, Carole Garbuny. *Shifting Shores (The Restless Sea).* London: Franklin Watts, 2003.

Voshell, J. Reese, Jr. *A Guide to Common Freshwater Invertebrates of North America.* Blacksburg, VA: McDonald and Woodward Publishing Company, 2002.

Woodward, Susan L. *Biomes of Earth: Terrestrial, Aquatic, and Human Dominated.* Westport, CT: Greenwood Press, 2003.

Worldwatch Institute, ed. *Vital Signs 2003: The Trends That Are Shaping Our Future.* New York: W.W. Norton, 2003.

Yahner, Richard H. *Eastern Deciduous Forest, Second Edition: Ecology and Wildlife Conservation.* Minneapolis, MN: University of Minnesota Press, 2000.

Zabel, Cynthia, and Robert G. Anthony. *Mammal Community Dynamics: Management and Conservation in the Coniferous Forests of Western North America.* New York: Cambridge University Press, 2003.

Periodicals

"Arctic Sea Ice Reaches Record Low." *Weatherwise.* 60. 6 Nov-Dec 2007: 11.

Bischof, Barbie. "Who's Watching Whom?" *Natural History.* 116. 10 December 2007: 72.

Bright, Chris, and Ashley Mattoon. "The Restoration of a Hotspot Begins." *World Watch* 14.6 Nov-Dec 2001: p. 8–9.

Carwardine, Mark. "So Long, and Thanks for All the Fish: If the Yangtze River Dolphin isn't Quite Extinct Yet, It Soon Will Be." *New Scientist.* 195. 2621 September 15, 2007: 50.

Coghlan, Andy. "Earth Suffers as We Gobble Up Resources." *New Scientist.* 195. 2611 July 7, 2007: 15.

Cunningham, A. "Going Native: Diverse Grassland Plants Edge Out Crops as Biofuel." *Science News.* 170. 24 December 9, 2006: 372.

Darack, Ed. "Death Valley Springs Alive." *Weatherwise.* 58. 4 July-August 2005: 42.

Darack, Ed. "The Hoh Rainforest." *Weatherwise.* 58. 6 Nov-Dec 2005: 20.

De Silva, José María Cardoso, and John M. Bates. "Biogeographic Patterns and Conservation in the South American Cerrado: A tropical Savanna Hotspot." *BioScience* 52.3 March 2002: p225.

El-Bagouri, Ismail H.M. "Interaction of Climate Change and Land Degradation: the Experience in the Arab Region." *UN Chronicle.* 44. 2 June 2007: 50.

Ehrhardt, Cheri M. "An Amphibious Assault." *Endangered Species Bulletin.* 28. 1 January-February 2003: 14.

Flicker, John. "Audubon view: Grassland Protection." *Audubon.* 107. 3 May-June 2005: 6.

Greer, Carrie A. "Your Local Desert Food and Drugstore." *Skipping Stones.* 20. 2 March-April 2008: 34.

Gunnard, Jessie, Andrew Wier and Lynn Margulis. "Mycological Maestros: in the Ecuadorean Rainforest, a 'Missing Link' in the Evolution of Termite Agriculture?" *Natural History.* 112. 4 May 2003: 22.

Haedrich, Richard L. "Deep Trouble: Fishermen Have Been Casting Their Nets into the Deep Sea After Exhausting Shallow-water Stocks. But Adaptations to Deepwater Living Make the Fishes There Particularly Vulnerable to Overfishing—and Many are Now Endangered." *Natural History.* 116. 8 October 2007: 28.

Hansen, Andrew J., et al. "Global Change in Forests: Responses of Species, Communities, and Biomes." *BioScience.* 51. 9 September 2001: 765.

Johnson, Dan. "Wetlands: Going, Goings … Gone?" *The Futurist.* 35. 5 September 2001: 6.

"Katrina Descends." *Weatherwise.* 58. 6 November-December 2005: 10.

Kessler, Rebecca. "Spring Back." *Natural History* 115.9 November 2006: p 18.

Kloor, Keith. "Fire (in the Sky): In Less than an Hour, Flames had Reduced Nearly 8,000 acres of Grasslands to Smoldering Stubble and Ash." *Audubon.* 105. 3 September 2003: 74.

Lancaster, Pat. "The Oman experience." *The Middle East.* 384 December 2007: 55.

Levy, Sharon. *New Scientist: Last Days of the Locust.* February 21, 2004, p. 48–49.

Marshall, Laurence A. "Sacred Sea: A Journey to Lake Baikal." *Natural History.* 116. 9 November 2007: 69.

Maynard, Barbara. "Fire in Ice: Natural Gas Locked Up in Methane Hydrates Could Be the World's Next Great Energy Source—If Engineers Can Figure Out How to Extract it Safely." *Popular Mechanics.* 183. 4 April 2006: 40.

Mohlenbrock, Robert H. "Wetted Bliss: in a Louisiana Refuge, Different Degrees of Moisture Create Distinctive Woods." *Natural History.* 117. 2 March 2008: 62.

Morris, John. "Will the Forest Survive?" *Wide World.* 13. 1 September 2001: 15.

Payette, Serge, Marie-Josee Fortin, and Isabelle Gamache. "The Subarctic Forest—Tundra: The Structure of a Biome in a Changing Climate." *BioScience.* 51. 9 September 2001: 709.

Pennisi, Elizabeth. "Neither Cold Nor Snow Stops Tundra Fungi." *Science.* 301. 5,638 September 5, 2003: 1,307.

Petty, Megan E. "The Colorado River's Dry Past." *Weatherwise.* 60. 4 (July-August 2007): 11.

Pollard, Simon D, and Robert R. Jackson. "Vampire Slayers of Lake Victoria: African Spiders get the Jump on Blood-filled Mosquitoes. (*Evarcha culicivora*)." *Natural History.* 116. 8 October 2007: 34.

Shefrin, Russell. "An Optimistic Look at Falling Leaves." *New York State Conservationist.* 62.2 October 2007: p32.

"Siberia Sees the Wood from the Trees." *Geographical.* 73. 1 January 2001: 10.

Springer, Craig. "Leading-edge Science for Imperiled, Bonytail." *Endangered Species Bulletin.* 27. 2 March-June 2002: 27.

Springer, Craig. "The Return of a Lake-dwelling Giant." *Endangered Species Bulletin.* 32. 1 February 2007: 10.

Sterling, Eleanor J, and Merry D. Camhi. "Sold Down the River: Dried Up, Dammed, Polluted, Overfished—Freshwater Habitats Around the World are Becoming Less and Less Hospitable to Wildlife." *Natural History.* 116. 9 November 2007: 40.

Stone, Roger D. "Tomorrow's Amazonia: as Farming, Ranching, and Logging Shrink the Globe's Great Rainforest, the Planet Heats Up." *The American Prospect.* 18. 9 September 2007: A2.

Tucker, Patrick. "Growth in Ocean-current Power Foreseen: Florida Team Seeks to Harness Gulf Stream." *The Futurist.* 41. 2 March-April 2007: 8.

Wagner, Cynthia G. "Battles On the Beaches." *The Futurist.* 35. 6 November 2001: 68.

Weir, Kirsten L. "Don't Tread on It." *Natural History.* 110. 9 November 2001: 37.

Yalden, Derek. "Managing moorland." *Biological Sciences Review.* 18. 2 November 2005: 14.

Organizations

African Wildlife Foundation, 1400 16th St. NW, Suite 120, Washington, DC 20036. Phone: 202-939-3333, Fax: 202-939-3332, Internet: http://www.awf.org.

American Cetacean Society, PO Box 1391, San Pedro, CA 90733. Internet: http://www.acsonline.org.

American Littoral Society, Sandy Hook, Highlands, NJ 07732. Phone: 732-291-0055, Internet: http://www.alsnyc.org.

American Oceans Campaign, 2501 M St., NW Suite 300, Washington DC 20037-1311. Phone: 202-833-3900. Fax: 202-833-2070, Internet: http://www.oceana.org.

American Rivers, 1101 14th St. NW, Suite 1400, Washington, DC 20005. Phone: 202-347-7550, Fax: 202-347-9240, Internet: http://www.americanrivers.org.

Canadian Lakes Loon Survey, PO Box 160, Port Rowan, ON, Canada N0E 1M0. Internet: http://www.bsc-eoc.org/cllsmain.html.

Center for Environmental Education, Center for Marine Conservation, 1725 De Sales St. NW, Suite 500, Washington, DC 20036.

Center for Marine Conservation, 1725 DeSales St., NW, Suite 600, Washington, DC 20036. Phone: 202-429-5609, Fax: 202-872-0619, Internet: http://www.cmc-ocean.org.

Chihuahuan Desert Research Institute, PO Box 905, Fort Davis, TX 79734. Phone: 432-364-2499, Fax: 432-364-2686, Internet: http://www.cdri.org.

Coast Alliance, PO Box 505, Sandy Hook, Highlands, NJ 07732. Phone: 732-291-0055, Internet: http://www.coastalliance.org.

Desert Protective Council, Inc., PO Box 3635, San Diego, CA 92163. Phone: 619-342-5524, Internet: http://www.dpcinc.org/environ_issues.shtml.

Envirolink, PO Box 8102, Pittsburgh, PA 15217. Internet: http://www.envirolink.org.

Environmental Defense Fund, 257 Park Ave. South, New York, NY 10010. Telephone: 212-505-2100, Fax: 212-505-2375, Internet: http://www.edf.org.

Environmental Network, 4618 Henry St., Pittsburgh, PA 15213. Internet: www.environlink.org

Environmental Protection Agency, 401 M St., SW, Washington DC 20460. Telephone:202-260-2090, Internet: http://www.epa.gov.

The Freshwater Society, 2500 Shadywood Rd., Navarre, MN 55331. Phone: 952-471-9773, Fax: 952-471-7685, Internet: http://www.freshwater.org.

Friends of the Earth, 1717 Massachusetts Ave. NW 300, Washington, DC 20036. Telephone:877-843-8687, Fax:202-783-0444, Internet: http://www.foe.org.

Global ReLeaf, American Forests, PO Box 2000, Washington, DC 20013. Telephone: 202-737-1944, Internet: http://www.amfor.org.

Global Rivers Environmental Education Network (GREEN), 2120 W. 33rd Avenue, Denver, CO 80211. Internet: http://www.earthforce.org/green.

Greenpeace USA, 702 H St. NW, Washington, DC 20001. Telephone:202-462-1177, Internet: http://www.greenpeace.org.

International Joint Commission, 1250 23rd St. NW, Suite 100, Washington, DC 20440. Phone: 202-736-9024, Fax: 202-467-0746, Internet: http://www.ijc.org.

Izaak Walton League of America, 707 Conservation Ln., Gaithersburg, MD 20878. Telephone: 301-548-0150; Internet: http://www.iwla.org

National Wetlands Conservation Project, The Nature Conservancy, 1800 N Kent St., Suite 800, Arlington, VA 22209. Phone: 800-628-6860, Internet: http://www.tnc.org.

Nature Conservancy, Worldwide Office, 4245 North Fairfax Dr., Arlington, VA 22203-1606. Phone: 800-628-6860, Internet: http://www.nature.org.

North American Lake Management Society, PO Box 5443, Madison, WI 53705-0443. Phone: 608-233-2836, Fax: 608-233-3186, Internet: http://www.nalms.org.

Olympic Coast Alliance, PO Box 573 Olympia, WA 98501. Phone: 360-705-1549, Internet: http://www.olympiccoast.org.

Project Wet, 1001 West Oak, Suite 210, Bozeman, MT 59717. Phone: 866-337-5486, Fax: 406-522-0394, Internet: http://projectwet.org.

Rainforest Alliance, 665 Broadway, Suite 500, New York, NY 10012. Phone: 888-MY-EARTH, Fax: 212-677-1900, Internet: http://www.rainforest-alliance.org.

Sierra Club, 85 St., 2nd Fl., San Francisco, CA 94105. Telephone: 415-977-5500, Fax: 415-977-5799, Internet: http://www.sierraclub.org.

The Wilderness Society, 1615 M St. NW, Washington, DC 20036. Telephone: 800-the-wild, Internet: http://www.wilderness.org.

World Meteorological Organization, 7-bis, avenue de la Paix, Case Postale No. 2300 CH-1211, PO Box 2300, Geneva 2, Switzerland. Phone: 41 22 7308111, Fax: 41 22 7308181, Internet: http://www.wmo.ch.

World Wildlife Fund, 1250 24th St. NW, Washington, DC 20090. Internet: http://www.wwf.org.

Web sites

"America's Wetlands." Environmental Protection Agency. http://www.epa.gov/OWOW/wetlands/vital/what.html (accessed July 13, 2007).

Arctic Studies Center.http://www.nmnh.si.edu/arctic (accessed August 9, 2007).

Blue Planet Biomes. http://www.blueplanetbiomes.org (accessed September 14, 2007).

CBC News Indepth: Oceans. http://www.cbc.ca/news/background/oceans/part2.html (accessed September 1, 2007).

Discover Magazine. http://www.discovermagazine.com (accessed September 12, 2007).

Distribution of Land and Water on the Planet. http://www.oceansatlas.com/unatlas/about/physicalandchemicalproperties/background/seemore1.html (accessed September 1, 2007).

Envirolink. http://www.envirolink.org (accessed September 14, 2007).

FAO Fisheries Department. http://www.fao.org/fi (accessed September 5, 2007).

"The Grassland Biome." University of California Museum of Paleontology. http://www.ucmp.berkeley.edu/exhibits/biomes/grasslands.php (accessed September 14, 2007).

Journey North Project. http://www.learner.org/jnorth (accessed August 14, 2007).

Long Term Ecological Research Network. http://lternet.edu (accessed August 9, 2007).

Monterey Bay Aquarium. http://www.mbayaq.org (accessed September 12, 2007).

National Center for Atmospheric Research. http://www.ncar.ucar.edu (accessed September 12, 2007).

National Geographic Magazine. http://www.nationalgeographic.com (accessed September 14, 2007).

National Oceanic and Atmospheric Administration. http://www.noaa.gov(accessed September 1, 2007).

National Park Service. http://www.nps.gov (accessed September 14, 2007).

National Park Service, Katmai National Park and Preserve. http://www.nps.gov/katm/ (accessed August 9, 2007).

National Science Foundation. http://www.nsf.gov (accessed August 9, 2007).

Nature Conservancy. http://www.nature.org (accessed September 5, 2007).

The Ocean Environment. http://www.oceansatlas.com/unatlas/about/physicaland chemicalproperties/background/oceanenvironment.html (accessed September 1, 2007).

Oceana. http://www.americanoceans.org (accessed August 14, 2007).

"On Peatlands and Peat." International Peat Society. http://www.peatsociety.fi (accessed July 13, 2007).

Ouje-Bougoumou Cree Nation. http://www.ouje.ca (accessed August 25, 2007).

Scientific American Magazine. http://www.sciam.com (accessed September 5, 2007).

"Status and Trends of Wetlands of the Counterminous United States from 1998 to 2004." U.S. Fish & Wildlife Service. http://wetlandsfws.er.usgs.gov/status_trends/National_Reports/trends_2005_report.pdf (accessed July 13, 2007).

Swedish Environmental Protection Agency. http://www.internat.naturvardsverket.se (accessed August 25, 2007).

Thurston High School: Biomes. http://ths.sps.lane.edu/biomes/index1.html (accessed August 9, 2007).

Time Magazine. http://time.com (accessed August 14, 2007).

"The Tundra Biome." University of California, Museum of Palentology. http://www.ucmpberkeley.edu/exhibits/biomes/tundra.php (accessed August 9, 2007).

U.N. Atlas of the Oceans. http://www.cmc-ocean.org (accessed September 12, 2007).

UNESCO. http://www.unesco.org/ (accessed September 5, 2007).

University of California at Berkeley. http://www.ucmp.berkeley.edu/glossary/gloss5/biome/index.html (accessed July 13, 2007).

U.S. Fish and Wildlife Services. http://www.fws.gov/

USDA Forest Service. http://www.fs.fed.us (accessed August 22, 2007).

"Wind Cave National Park." The National Park Service. http://www.nps.gov/wica/ (accessed September 14, 2007).

Woods Hole Oceanographic Institution. http://www.whoi.edu (accessed May 14, 2008).

The World Conservation Union. http://www.iucn.org (accessed August 22, 2007).

World Wildlife Fund. http://www.wwf.org (accessed August 22, 2007)

World Meteorological Organization. http://www.wmo.ch (accessed August 17, 2007).

Index

Numerals in italic type indicate volume number. Bold type indicates a main entry. Graphic elements (photographs, tables, illustrations) are denoted by (ill.).

A

Aborigines (Australia)
    clothing, *1:* 156
    Kakadu region, *3:* 503
    overview, *1:* 164
    tracking abilities, *1:* 159
Abyssal plains, *2:* 268
Abyssopelagic zones, *2:* 259
Acacia trees, *1:* 98, 110; *2:* 185, 185 (ill.), 188
Academy of Applied Science, *2:* 234
Acadia National Park, *3:* 419–20
Accidents
    plane crashes, *1:* 129, 159
    shipwrecks, *2:* 286
Acid rain
    coniferous forests, *1:* 36
    deciduous forests, *1:* 113
    lakes and ponds, *2:* 242
    wetlands, *3:* 494–95
Acornic acid, *1:* 112
Acorns, *1:* 112
Acquired immune deficiency syndrome (AIDS), *2:* 326
Active continental margins, *1:* 55
Active layers (tundra soil), *3:* 430–31
Addaxes, *1:* 153
"Adopt a Wetland" programs, *3:* 495
Africa
    grassland dwellers, *2:* 199–200
    hardwood trees, *1:* 113*t*

rain forests, *2:* 331–32
savannas, *2:* 175, 176
softwood trees, *1:* 7*t*
See also specific countries
African catfish, *3:* 361
African lungfish, *3:* 487
The African Queen (film), *3:* 370
African running frogs, *2:* 190
African weaverbirds, *2:* 192
Agama lizards, *1:* 27 (ill.), 143; *2:* 190
Age
    rivers and streams, *3:* 339–40, 340 (ill.)
    trees, *1:* 11, 96; *2:* 310
Agent Orange, *3:* 501–2
Agriculture. *See* Farming and farmland
Agulhas Current, *2:* 290
AIDS (acquired immune deficiency syndrome), *2:* 326
Ainu people, *1:* 114–15, 115 (ill.)
Air quality and pollution
    deciduous forests, *1:* 112, 113
    rain forests, *2:* 328
    tundra regions, *3:* 451–52
    See also Acid rain
Air temperatures
    boreal forests, *1:* 9
    coniferous forests, *1:* 10
    deserts, *1:* 124, 126–27
    grasslands, *2:* 179
    hurricanes and typhoons, *2:* 261
    ocean effect on, *2:* 258, 264

M

T